BLOCKCHAIN-BASED DIGITAL TWINS

Research Trends and Challenges

BLOCKCHAIN-BASED DIGITAL TWINS

Research Trends and Challenges

A. Prasanth, PhD
Vandana Sharma, PhD
Rajesh Kumar Dhanaraj, PhD
Pethuru Raj, PhD
Editors

AAP | APPLE ACADEMIC PRESS

First edition published 2025

Apple Academic Press Inc.
1265 Goldenrod Circle, NE,
Palm Bay, FL 32905 USA

760 Laurentian Drive, Unit 19,
Burlington, ON L7N 0A4, CANADA

CRC Press
2385 NW Executive Center Drive,
Suite 320, Boca Raton FL 33431

4 Park Square, Milton Park,
Abingdon, Oxon, OX14 4RN UK

Library and Archives Canada Cataloguing in Publication

CIP data on file with Canada Library and Archives

Library of Congress Cataloging-in-Publication Data

CIP data on file with US Library of Congress

ISBN: 978-1-77491-842-5 (hbk)
ISBN: 978-1-77491-843-2 (pbk)
ISBN: 978-1-00356-904-6 (ebk)

About the Editors

A. Prasanth, PhD
Associate Professor, Department of Computer Science and Engineering, Vel Tech Rangarajan Dr. Sagunthala R&D Institute of Science and Technology, Chennai, India

A. Prasanth, PhD, is currently working as an Associate Professor in the Department of Computer Science and Engineering at Vel Tech Rangarajan Dr. Sagunthala R&D Institute of Science and Technology, Chennai, Tamil Nadu, India. He has published more than 40 research articles in international journals and conference proceedings, among which 15 were indexed in SCI and 22 were indexed in Scopus. In addition, he has published eight patents and 10 books under a reputed publisher. He has served as a recognized Anna University PhD supervisor with four scholars conducting their research under his guidance. One scholar has completed his PhD under Professor Prasanth. His research interests include the Internet of Things, blockchain, machine learning, wireless sensor networks, ad hoc networks, and computer networks. He received his BE in Electronics and Communication Engineering, ME in Computer Science and Engineering (with a specialization in Networks), and his PhD degree in Information and Communication Engineering, all from Anna University, Chennai, India.

Vandana Sharma, PhD
Associate Professor, CHRIST (Deemed to be University), Delhi NCR, India

Vandana Sharma, PhD, is an Associate Professor at CHRIST (Deemed to be University), Delhi NCR, India. Dr. Sharma has served at NAAC-accredited A++ institutions such as Guru Gobind Singh Indraprastha University, Delhi; Galgotias University, Greater Noida Campus; and Amity Institute of Information Technology, Amity University Noida Campus, India. She is a senior member of IEEE, a member of the Women in Engineering Society, and a member of the IEEE Consumer Technology Society, popularly known as CTSoc. As a keen researcher, she has published more than 50 research papers in SCI- and Scopus-indexed international journals and conferences. Dr. Sharma has contributed as a keynote speaker, session chair, reviewer, and technical program committee member for reputed international journals and IEEE conferences and has presented her work across India and abroad. Currently, she is a postdoctoral fellow at Lincoln University College, Malaysia. Additionally, she is an honorary Adjunct Professor and Senior Fellow of the Scientific Innovation Research Group at Beni Suef University, Egypt, and an Adjunct Professor at Perdana University, Malaysia. Her primary areas of interest include artificial intelligence, blockchain technology, and the Internet of Things (IoT). Dr. Sharma earned her doctoral degree at Amity University and her Master of Technology and Bachelor of Electronic Commerce degrees from Guru Gobind Singh Indraprastha University, a Delhi state university in India.

Rajesh Kumar Dhanaraj, PhD

Professor, Symbiosis International (Deemed University), Pune, India

Rajesh Kumar Dhanaraj, PhD, is a Professor at the Symbiosis Institute of Computer Studies and Research, Symbiosis International (Deemed University), Pune, India. He has contributed to more than 25 authored and edited books on various technologies, over 50 articles and papers in various refereed journals and international conferences, as well as book chapters. He holds 21 patents. His research interests include machine learning, cyber-physical systems, and wireless sensor networks. He is a senior member of the Institute of Electrical and Electronics Engineers and a member of the Computer Science Teachers Association and the International Association of Engineers. He is an expert advisory panel member of Texas Instruments Inc., USA. He received his BE degree in Computer Science and Engineering and his MTech and PhD degrees in Computer Science, all from Anna University, Chennai, India.

Pethuru Raj, PhD

Chief Architect and Vice President, Reliance Jio Platforms Ltd. (JPL), Bangalore, India

Pethuru Raj, PhD, is Chief Architect and Vice President at Reliance Jio Platforms Ltd. (JPL), Bangalore, India. Previously, he worked as a cloud architect at the IBM Global Cloud Center of Excellence (CoE), as an enterprise architect in Wipro Consulting Services (WCS), and as a software architect in Robert Bosch Corporate Research. In total, he has gained more than 22 years of IT industry experience and eight years of research experience. He finished his CSIR-sponsored PhD degree at Anna University, Chennai, and continued with his UGC-sponsored postdoctoral research in the Department of Computer Science and Automation at the Indian Institute of Science, Bangalore. He was granted a couple of international research fellowships (JSPS and JST) to work as a research scientist for several years at two leading Japanese universities. He has authored and edited more than 45 technology books, published more than 60 research papers, and contributed more than 50 book chapters. Dr. Raj has also submitted a couple of international patents. He completed several IT architecture certifications (TOGAF and ITIL) as well. He has delivered keynote addresses at 25 international conferences and has conducted numerous technical sessions across colleges, universities, and institutes of eminence in India and abroad. He has participated as a panel member in IT industry symposiums and written several white papers for IT companies.

Contents

Contributors

E. A. Mary Anita
CHRIST (Deemed to be University), Bangalore, Karnataka, India

T. Arunkumar
Department of Computer Science Engineering, Kongu Engineering College, Perundurai, Tamil Nadu, India

Arun Seeralan Balakrishnan
Department of Mechatronics Engineering, Asia Pacific University of Technology and Innovation, Kuala Lumpur, Malaysia

A. Bharathi
Assistant Professor and Research Scholar, Department of Information Technology, School of Computing Sciences, VISTAS Pallavaram, Chennai, Tamil Nadu, India

Mani Deepak Choudhry
Department of Computing Technologies, SRM Institute of Science and Technology, Kattakulathur, Tamil Nadu, India

P. V. Deepa
Department of Electronics and Communication Engineering, Arunachala College of Engineering for Women, Kanyakumari, Tamil Nadu, India

G. Gayathri Devi
Department of Computer Science, Shrimathi Devkunvar Nanalal Bhatt Vaishnav College for Women, Chennai, Tamil Nadu, India

M. Parimala Devi
Department of Electronics and Communication Engineering, Velalar College of Engineering and Technology, Tamil Nadu, India

Rajesh Kumar Dhanaraj
Professor, Symbiosis International (Deemed University), Pune, Maharashtra, India

J. Dharani
Department of Information Technology, Dr. N.G.P. Institute of Technology, Coimbatore, Tamil Nadu, India

Ahmed A. Elngar
Faculty of Computers and Artificial Intelligence, Beni-Suef University, Egypt

P. Geetha
Department of Computational Intelligence, SRM Institute of Science and Technology, Kattakulathur, Tamil Nadu, India

R. D. Gomathi
Department of English, Kongu Engineering College, Perundurai, Tamil Nadu, India

U. Hemamalini
Assistant Professor and Research Scholar, Department of Information Technology, School of Computing Sciences, VISTAS Pallavaram, Chennai, Tamil Nadu, India

S. Jeevanandham
Department of Information Technology, KGiSL Institute of Technology, Coimbatore, Tamil Nadu, India

Akshya Jothi
Department of Computational Intelligence, SRM Institute of Science and Technology, Kattankalathur, Chennai, Tamil Nadu, India

Seifedine Kadry
Department of Applied Data Science, Noroff University College, Kristiansand, Norway; Artificial Intelligence Research Center (AIRC), Ajman University, Ajman, United Arab Emirates; Department of Electrical and Computer Engineering, Lebanese American University, Byblos, Lebanon

V. Karpagam
Department of Artificial Intelligence and Data Science, Sri Ramakrishna Engineering College, Coimbatore, Tamil Nadu, India

S. Karthik
Department of Computer Science and Engineering, SNS College of Technology, Coimbatore, Tamil Nadu, India

Gaganjot Kaur
Project Manager, Iotasol Technologies, Ludhiana, Punjab, India

M. S. Kavitha
Department of Computer Science and Engineering, SNS College of Technology, Coimbatore, Tamil Nadu, India

P. V. Kavitha
Department of Artificial Intelligence and Data Science, Sri Ramakrishna Engineering College, Coimbatore, Tamil Nadu, India

A. Mohan Kumar
Tata Consultancy Private Limited, Chennai, Tamil Nadu, India

S. Jothi Lakshmi
Associate Professor, Department of Computer Science and Engineering, Akshaya College of Engineering and Technology, Coimbatore, Tamil Nadu, India

S. Maheswaran
Department of Electronics and Communication Engineering, Kongu Engineering College, Perundurai, Tamil Nadu, India

P. Manjula
Associate Professor, Department of Computer Science and Engineering, Saveetha School of Engineering, Chennai, Tamil Nadu, India

Sugandhi Midha
Assistant General Manager, College Dekho, Chandigarh, India

S. Nandita
Department of Electronics and Communication Engineering, Kongu Engineering College, Perundurai, Tamil Nadu, India

Mariya Ouaissa
Professor, Moulay Ismail University, Meknes, Morocco

Chien Thang Pham
TNU–University of Sciences, Thai Nguyen, Vietnam

R. Pradeep
Assistant Professor, Department of Electronics and Communication Engineering, Sri Ramakrishna Engineering College, Coimbatore, Tamil Nadu, India

A. Prasanth
Associate Professor, Department of Computer Science and Engineering, Vel Tech Rangarajan Dr. Sagunthala R&D Institute of Science and Technology, Chennai, India

R. Radha
Department of Computer Science, Shrimathi Devkunvar Nanalal Bhatt Vaishnav College for Women, Chennai, Tamil Nadu, India

B. Ragavi
Department of Biomedical Engineering, Dr. N.G.P. Institute of Technology, Coimbatore, Tamil Nadu, India

G. Boopathi Raja
Department of Electronics and Communication Engineering, Velalar College of Engineering and Technology, Tamil Nadu, India

P. Rithika
Department of Electronics and Communication Engineering, Kongu Engineering College, Perundurai, Tamil Nadu, India

S. Roobini
Department of Computer Science and Engineering, SNS College of Technology, Coimbatore, Tamil Nadu, India

T. Sathya
Department of Electronics and Communication Engineering, Velalar College of Engineering and Technology, Tamil Nadu, India

I. Mohammed Shafiq
Department of Electronics and Communication Engineering, Kongu Engineering College, Perundurai, Tamil Nadu, India

S. Sharmila
Department of Biomedical Engineering, Dr. N.G.P. Institute of Technology, Coimbatore, Tamil Nadu, India

K. K. Devi Sowndarya
Assistant Professor, Department of Electronics and Communication Engineering, DMI College of Engineering, Chennai, Tamil Nadu, India

M. Sundarrajan
Department of Computer Science and Engineering, SRM Institute of Science and Technology, Ramapuram Campus, Chennai, Tamil Nadu, India

Padmavathy Thirunavukkarasu
Department of Computer Science and Engineering, Sri Venkateswara College of Engineering, Sriperumbudur, Tamil Nadu, India

K. Veena
Assistant Professor, Computer Science and Engineering, Akshaya College of Engineering and Technology, Coimbatore, Tamil Nadu, India

D. Vinodha
CHRIST (Deemed to be University), Bangalore, Karnataka, India

Preface

Blockchain technology has leveraged most of the manufacturing and service industries by deploying inviolable trust. It is the design, the blueprint of a product, to the sales and after-sale agreements for the product. All transactions are documented well in this distributed ledger. Blockchain has identified many applications across industries, such as integrated architecture for handling IoT ecosystems, measuring diagnosis accuracy, blockchain-based smart tracing and tracking platforms for supply chains, and handling privacy and security preservation of health records.

Huge growth in the usage of blockchain, with the amalgamation of AI and big data, has exploited the potential of the data generated from current industrial practices. The trustworthiness of blockchain technology and its digital twin has greater reach when compared to traditional systems. The blockchain will intelligently automate the generation of smart contracts from the Industrial Internet of Things, parallel governance, cybersecurity, and the healthcare sector in providing policy-making processes.

Today, the parallel existence of blockchain-based digital twins influences AI, ML, and IoT to integrate smart contracts across industries. In the real world, document frauds are very common; thus, blockchain-based smart contracts will help mitigate this problem and assist in easy record-keeping from the beginning of the contract until the end. The smart contract will be viable to assist in legal proceedings.

The world is keen to leverage multi-faceted blockchain techniques to deploy and deliver newer business models. Recent times have witnessed tremendous growth in blockchain technology, its related framework, and the coexistence of digital twins. These come in handy in replicating the physical world with cyber-physical space. Resource-intensive service models spread across a variety of environments (including smart contracts, smart supply chain management, etc.) are meticulously empowered with blockchain-based digital twins. These facilitate blockchain-based digital twin processes to join the mainstream world. All kinds of connected products are collectively and/or individually enabled to be secure in their operations, offerings, and outputs. The edited book on blockchain-based digital twins addresses the next-generation technology to actualize secure systems, networks, and environments.

In the case of smart contract execution in cyberspace, the use of digital twin technology will help in fixing the authority and responsibility of the parties involved. The research trends for blockchain-based digital twins in Agriculture 4.0, supply chain management, and smart logistics are evolving, and the future directions for a suitable policy and framework are the need of the hour.

CHAPTER 1

Introduction to Transformation of Blockchain-Based Digital Twins

R. RADHA,[1] G. GAYATHRI DEVI,[1] RAJESH KUMAR DHANARAJ,[2] and AHMED A. ELNGAR[3]

[1]*Department of Computer Science, Shrimathi Devkunvar Nanalal Bhatt Vaishnav College for Women, Chennai, Tamil Nadu, India*

[2]*Professor, Symbiosis International (Deemed University), Pune, Maharashtra, India*

[3]*Faculty of Computers and Artificial Intelligence, Beni-Suef University, Egypt*

ABSTRACT

The concept of blockchain was first introduced in October 2008, as a means of providing security features for Bitcoin transactions without the need for third-party intermediaries like banks. Blockchain technology operates on a peer-to-peer network, a distributed architecture that connects computing devices without a central server. Decentralization of data pooling, data immutability, and one-to-one communication between computers or Internet of things (IoT) devices are the key concepts of blockchain. Digital twins (DT) are virtual replicas of physical objects or systems, created using data and information from sensors, IoT devices, and other sources. By simulating the behavior and performance of real-world objects or systems, DTs can help organizations to better understand and optimize their operations, identify potential problems before they occur, and develop more effective maintenance and repair

Blockchain-Based Digital Twins: Research Trends and Challenges.
A. Prasanth, Vandana Sharma, Rajesh Kumar Dhanaraj, Pethuru Raj (Eds.)

strategies. DTs are also used to support designers and engineers to test and refine their designs before they are physically built. When blockchain and DT technologies are combined they bring several benefits to various industries, especially those that rely on complex supply chains or critical infrastructure. Potential applications of these two technologies are supply chain management, asset management, smart contracts, and data sharing in various fields where security, privacy, transparency, and efficient management are the major concerns. The application where these two technologies are amalgamated is discussed in this chapter.

1.1 INTRODUCTION

The first Industrial Revolution, also known as Industry 1.0, occurred in the 17th century and involved the introduction of large-scale manufacturing, aided by coal and steam as energy sources. This brought about a significant increase in productivity and the replacement of manual labor with machinery such as looms, gears, steam engines, and ships. The second Industrial Revolution, Industry 2.0, took place in the 19th century with the discovery of electricity and oil fields, resulting in the development of the assembly line for streamlining production [1].

Industry 3.0 emerged around 1970 in the 21st century with the introduction of automation in factories through the use of computers to replace repetitive tasks. We are now in the era of Industry 4.0, the Fourth Industrial Revolution, which incorporates Cyber-Physical Systems (CPS), smart sensors, and data storage to enable production systems to communicate and share information [2].

Industry 4.0 is a term used by industry professionals to describe the integration of new technologies such as IoT [3–5], AI, VR, Cloud Computing (CC), and blockchain into their production processes. These technologies are incorporated into smart factories that feature embedded software, advanced sensors, and robotics to collect and analyze data for decision-making. Businesses can also benefit from better logistics, supply chain tracking, and improved customer service. By implementing AI-powered sensors and devices, manufacturing errors can be reduced and time saved. With a smartphone, a person can monitor the operations of their manufacturing unit with minimal investment. The application of machine learning algorithms can detect errors immediately, preventing costly issues down the line.

Industry 4.0 technology is being implemented in many fields, from agriculture to medicine. This chapter will discuss the transformation from Industry 3.0 to Industry 4.0 in several industries, highlighting the use of blockchain-based DTs technology. The morphological analysis framework, which helps to come across and discover research gaps in the field of quality management models, is explained. Developing more connections automating the customer market and the use of Lean Six Sigma in solving complex business problems can be done with the help of statistics [6].

1.1.1 BLOCKCHAIN

In a centralized network, all nodes are connected to a single server. If the server fails, the operations of the other nodes will be paused, and no processing of information can take place as all the nodes depend on the central server for data. In contrast, in a decentralized network, there is no concept of a single server that monitors every other node. Each node in the network has its entity and contributes to the network. Even if a problem occurs in one node, it will not affect the flow of information and decision-making, as the other nodes in the network can continue to function. Since blockchain technology is a decentralized network, decision-making requires collaboration and consensus with other nodes [7].

The blockchain can be divided into three primary components: a distributed network, a shared ledger, and digital transactions:

1. **Distributed Network:** The distributed network is made up of individual decision-making systems connected in a peer-to-peer network with nodes comprising network participants. Each member has an identical copy of the transactions and shares their information as a block after validating the data.

2. **Shared Ledger:** This records the digital transactions of the users as blocks. In a shared or distributed ledger system, every transaction is shared and has a copy in each node of the peer-to-peer network. It provides a universal state layer and data that enables every actor on the internet, irrespective of different geographical locations and jurisdictions, to interact over the internet without

the need for third parties like banks and to avoid double-spending problems. Any change to the shared ledger will be carried over to all copies that exist in the blockchain network. Once a transaction is completed, it becomes immutable, and data cannot be tampered with or deleted by any single member in the network because every member has a copy of the complete transactions.

3. **Digital Transactions:** These are valid information that is digitally saved in the network in an encrypted form and is authenticated by the digital signature of the person who added it. Transactions are added as blocks, and each block has a hashed cryptographic function that identifies the previous block in the blockchain.

The blockchain is categorized into two types: private and public. A public blockchain network has no restrictions on the number of people who can join and share information, whereas a private blockchain-implemented network has restrictions exists in a confined place, and is controlled by a particular organization or entity.

1.1.1.1 BASIC OPERATIONS OF BLOCKCHAIN

Figure 1.1 shows the basic architecture of blockchain where one block consists of various types of information such as network transactions, security-related documents, and product prototypes. In blockchain, the first block will be called a genesis block which means it has no previous blocks. When new information is added, it becomes a block with a reference to the previous block, forming a chain of immutable information. These chains of blocks are saved as digital ledgers across the network, which get updated with every new block added to the chain. Since the blockchain is organized in a decentralized manner, even if one block fails, the information is still available in other blocks, thus ensuring no loss of data. The blockchain appends new or updated information as blocks to the existing blockchain, with the history of the previous block being recorded and unalterable. Therefore, records cannot be tampered with [8].

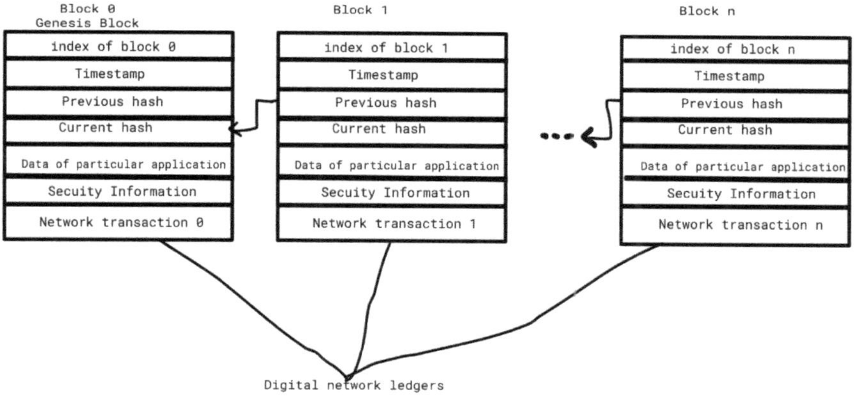

FIGURE 1.1 The basic architecture of blockchain.

1.1.1.2 BLOCKCHAIN WITH IOT

Blockchain technology has emerged as a potential solution for securing the growing volumes of real-time data generated by IoT devices and other IoT devices. With its decentralized and tamper-proof architecture, the blockchain offers a high level of security and transparency, which is crucial for ensuring the integrity and authenticity of the data. Additionally, the blockchain's ability to support smart contracts enables the automation of various processes and facilitates the efficient exchange of value, thereby enhancing the overall efficiency and effectiveness of the IoT systems. As a result, the blockchain is increasingly being integrated with IoT devices and other IoT devices to enable secure and intelligent real-time data exchange and processing [9].

Blockchain technology has the potential to bring significant benefits to various industries, including healthcare, retail, disaster management, and smart city planning. In the healthcare industry, for example, blockchain can be used to securely store and share patient data among healthcare providers, leading to better diagnosis, treatment, and care for patients. In the retail industry, blockchain can be used to create a more personalized customer experience by securely tracking and sharing customer data. In disaster management, blockchain can be used to provide an immutable and transparent record of relief efforts, ensuring that resources are being allocated effectively. And in smart city planning, blockchain can be used to securely store and share data on energy usage, transportation, and other infrastructure, leading to more efficient and sustainable cities.

Blockchain technology has the potential to reshape the future by providing secure, transparent, and decentralized solutions for various industries. Its immutability and programmability make it a more apt technology for a wide range of applications in the fields of finance, supply chain management, healthcare, and more. It eliminates the need for trusted intermediaries, allowing for a peer-to-peer network where participants can directly interact with each other without the need for intermediaries like banks, brokers, or governments. This not only reduces transaction costs but also provides greater transparency and security for the participants. As blockchain technology continues to evolve, it will likely become more integrated into our everyday lives and become a critical component of global business.

1.1.2 DIGITAL TWINS

With the advent of DTs, the simulation applications are more enhanced and experimentally advantageous as experimental DTs communicate with each other in the virtual world. By creating a virtual representation of a physical object and its environment, a DT provides a path for experimentation and testing in a controlled environment without the need for costly physical prototypes or risking damage to real-life objects. This enables more rapid and efficient development, optimization, and maintenance of various types of systems and processes. Additionally, the ability for DTs to communicate and collaborate with other DTs, both virtually and in the physical world, opens up more new possibilities for advanced modeling, simulation, and optimization of complex systems [10].

DTs allow for testing and analyzing the behavior of a physical object in a virtual environment before it is built in the real world. By receiving feedback from sensors and simulating various scenarios, DTs can predict the performance and potential problems of a product, which helps in reducing the time and cost of development. Additionally, the integration of various technologies like IoT, AI, and CC makes DTs more powerful and versatile in predicting and analyzing the behavior of physical objects as given in Figure 1.2.

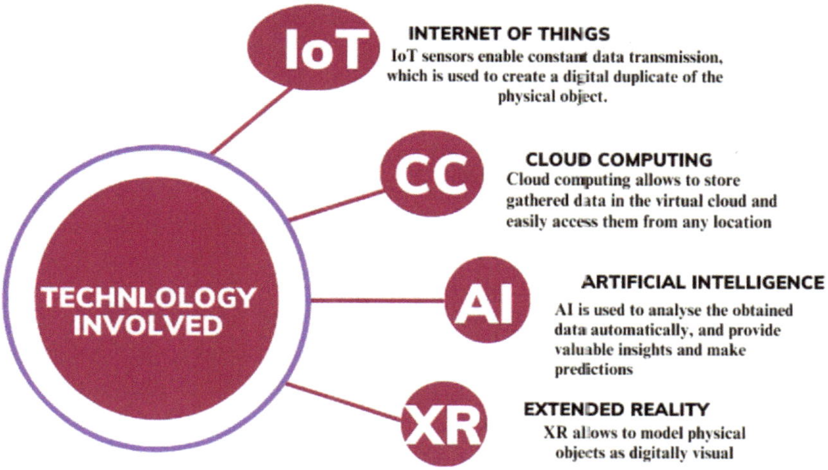

FIGURE 1.2 Technologies involved.

1. **IoT:** It is a key technology in the development and operation of DTs as it provides the means to collect real-time data from sensors attached to physical objects [11]. This data is then used to create a digital replica or twin of the physical object. The DT is then used to simulate and analyze the performance of the physical object in various conditions and to identify any potential issues before their occurrence in the real world.
2. **Extended Reality (XR):** The data collected from IoT sensors is used to create a DT, which is a virtual representation of a physical object or system. This DT can be visualized in various ways, including 3D and 4D (with time as the fourth dimension) visualizations, allowing for a better understanding of the physical object and its behavior.
3. **Cloud Technology (CC):** It plays a crucial role in DT technology, as it provides the necessary infrastructure for storing and processing the vast amounts of data collected by IoT sensors. The data can be stored in cloud-based servers, and DTs can be created using this data, which can then be accessed from anywhere in the world with an internet connection. This allows for easy collaboration and sharing of data, and insights, and also enables remote monitoring and management of physical objects.

4. **AI:** DTs are used to build simulations that replicate the behavior of physical objects in real life, which can then be analyzed and used to make informed decisions. The ability to simulate and predict how real-world objects will behave in different scenarios can be incredibly valuable in fields like engineering, manufacturing, and healthcare. By using DTs to model and test different scenarios, organizations can make more informed decisions and reduce the risk of costly errors or mistakes.

The concept of DTs was introduced by David Gelernter in his book Mirror Worlds in 1991. Later, in 2002, Michael Grieves from the University of Michigan coined the term "DT" and applied it to manufacturing. NASA then adopted the concept in 2010 to create digital simulations of spacecraft and capsules for testing before launching them into space. Since then, the use of DTs has expanded to various fields and industries.

The data collected by IoT components is stored in the DT using appropriate data structures and user interfaces to ensure that the data is stored in an organized and usable format. This helps in simulating and analyzing the behavior of the physical entity in various conditions and predicting its performance. Digital footprints are the traces left behind by a person's online activities and interactions, such as browsing history, social media posts, online purchases, and more. These footprints can be used to gather information about a person's interests, behaviors, preferences, and even location.

This collected information can then be used to personalize advertising, recommend products, and more. Individuals need to be aware of their digital footprints and take steps to protect their privacy online. It is possible to use the data collected from an individual's digital footprint to make predictions about their health and potential risk for certain diseases. However, it is important to note that this type of predictive analysis is not without ethical considerations, such as privacy concerns and the potential for biases in the data and algorithms used. Additionally, it is important to recognize that the predictions made by a DT are not foolproof and should always be validated through traditional medical testing and diagnoses.

Mobile devices and smartwatches with an internet connection can collect data related to a person's movement, activities, and health details such as sleeping patterns, blood pressure, weight, heart rate, workout hours, drinking water measurement, blood sugar level, and many more. These parameters can then be used as inputs to the DT to build a comprehensive model of the person's health and predict any potential health risks or issues. It is important to note that the use of DTs for criminal identification raises serious ethical and legal concerns. While it is true that data collected about a person's activities can be used to create a DT and potentially predict future behavior, the use of such technology for law enforcement purposes must be balanced against individuals' rights to privacy and due process.

In many countries, laws and regulations have been put in place to protect individuals' personal information and prevent its misuse by government authorities. The single-user and multi-user activities and how to record their activities digitally were discussed in Ref. [12]. The collaboration of humans and machines in the field of intelligent manufacturing was discussed, and the study was based on visual question answering applied with DTs that increase efficiency.

DTs can help in identifying the strengths and weaknesses of physical objects, and this information can be used to improve their performance. By creating a digital replica of a physical object, companies can run simulations and test new features or designs without actually having to build them in the real world. This can help to reduce costs and risks associated with physical experimentation and can lead to more efficient and effective product development. The use of DTs, which involve creating precise virtual replicas of real-world objects and simulating their operational processes, empowers businesses across various industries to explore innovative opportunities and improve their operations and outcomes with excellence. One approach is to start by implementing these systems in DTs, which can help identify drawbacks and improve the efficiency and robustness of the physical system. This way, businesses can detect both risks and opportunities before investing in building a complex and highly integrated system from scratch [13].

One of the key benefits of digitization is the ability to store historical and current information updates. This digital information is then used to determine our preferences, as seen in the recommendation systems used by various applications such as Netflix. Nowadays, the success of a business often hinges on its ability to collect digital data, which can then be used to display targeted advertisements to customers based on their interests and previous behavior. For instance, if a customer views a particular product on Facebook or Amazon, similar products may be recommended to them through advertisements, ultimately enticing them to make a purchase.

The technologies used with DTs their challenges and the avenue for more research in this field were discussed [14].

The use of DTs is not limited to specific industries but spans across various fields. For example, an electrical engineer designing an electronic circuit and a dentist planning orthodontic treatment use DTs through scans and simulations before implementation. With the availability of cloud technology that provides extensive computational power and unlimited storage, exciting opportunities arise for DTs [15].

1.1.3 ADVANTAGE OF AMALGAMATION OF DATA ANALYSIS, BLOCKCHAIN, DTS TECHNOLOGIES

1. **Data Integrity:** Blockchain technology guarantees data integrity and transparency by using encrypted data transfer, which also enables the traceability of transactions.
2. **Data Security:** DTs are stored as blocks in all nodes of a blockchain system, making it nearly impossible for hackers to manipulate or hack data without finding and compromising all instances of the DTs. Additionally, if any attempt to hack or modify data is made, it will be immediately visible to all other nodes, and the corrupted block will be rejected, ensuring that the blockchain system remains secure.
3. **Real-Time Data Analytics and Predictions:** The data and behavior of DTs can be analyzed for prediction of the behavior of the system, streamlining many processes.
4. **Data Sharing:** Enables project teams to avoid using data that has already been utilized, preventing duplication of data analytics.

The combination of DT with blockchain removes any loopholes and makes the system more trustworthy for users for their everyday operations.

1.2 APPLICATIONS OF AMALGAMATED DT AND BLOCKCHAIN

1.2.1 DT IN AUTOMOTIVE AND AEROSPACE

By creating a digital twin of a machine, it becomes possible to monitor its performance in real-time and collect data about its usage and maintenance history. This information is used to make predictions about the machine's remaining lifespan and the optimal time to replace its parts. This process is known as predictive maintenance and can help businesses save time and money by avoiding unexpected downtime and reducing maintenance costs.

By creating a digital twin of a physical object and collecting data on its behavior and performance over time, it becomes possible to predict when maintenance will be required, as well as to identify potential areas for improvement and optimization. This can lead to significant cost savings, improved efficiency, and increased lifespan for the physical object.

Digital twin technology is widely used in aircraft and vehicle management for various purposes such as fuel filling, tracking, weather prediction, weight monitoring, and detecting defects. By creating a virtual model of the aircraft or vehicle, digital twin technology can be used to monitor and analyze the performance of various components in real time and predict maintenance needs or potential issues before they occur. This helps improve the efficiency and safety of aircraft and vehicles and can also save costs by avoiding unexpected downtime or repairs [16].

1.2.2 DTS AND BLOCKCHAIN IN INDUSTRY

Industry 4.0 technologies have revolutionized data collection in various industries, providing benefits and optimization. By implementing DT and IoT technologies, the supply chain can have access to more critical and secure data, mitigating the risk of trusting external parties with their Intellectual Property Rights (IPR). DT technology can be used to create a virtual model of a physical product before it is built. This model can be tested and improved in great detail, enabling the identification of any potential problems that may arise in all possible situations. By using this technology, businesses can identify and resolve issues before manufacturing or finalizing the product, ensuring a high-quality outcome [17].

DT technology enables product development teams to understand the impact of improving the product in its later versions and anticipate the product output for future years, thereby allowing for the provision and planning for its continued improvement. Creating a DT of a product on a blockchain network enables the storage of information about the product and its manufacturers. Manufacturing companies can then track their product supply chain globally, which can help eliminate the theft of goods across the globe.

Counterfeit products pose a significant problem in today's market, as they can harm a company's intellectual property rights and affect stakeholders' investments, brand value, and product reputation. Therefore, companies must take measures to ensure that the products they manufacture are not counterfeit. The combination of blockchain and DT can help overcome the problem of counterfeit products. By introducing DTs during product development, companies can predict the success and failure of the product. When this technology is integrated with blockchain, it ensures the product's originality. Blockchain technology provides data transparency and enables the creation of digital certificates, while DT technology ensures the digitization of products and assets, providing a more secure and reliable system.

The amalgamation of blockchain and DTs can enhance the credibility of data and lead to more efficient and accurate decision-making in real-time applications such as logistics and the utility industry, resulting in improved productivity [18]. Today, almost all entities, such as homes, hotels, hospitals, and public and private communication systems, have undergone digital transformation. They generate vast amounts of data that are stored in the cloud and used to optimize their operations through the use of AI algorithms and robust optimization algorithms. As technology advances, consumers are becoming increasingly sophisticated and are seeking more customization to meet their individual needs. This trend is driving manufacturers to implement more customizable features in their products and become increasingly competitive in the marketplace [19]

1.2.3 DT IN SMART CITIES

1.2.3.1 MANAGING TRAFFIC

Traffic congestion is a major problem in urban cities and can have a significant impact on the economy and quality of life. It can cause delays in transportation, increase fuel consumption, and air pollution, and decrease productivity. Smart transportation systems, which utilize IoT and other technologies, can help alleviate traffic congestion by providing real-time traffic information, optimizing traffic signals, and implementing dynamic tolling systems to manage traffic flow. By analyzing traffic data and making intelligent decisions based on that data, cities can create more efficient and sustainable transportation networks [20].

Autonomous vehicles augmented with AI can potentially solve the problem of traffic congestion by enabling efficient, safe, and smart transportation. With the use of sensors, cameras, and advanced algorithms, autonomous vehicles can monitor and adapt to traffic patterns in real-time, avoiding traffic congestion by choosing the optimal route and speed [21]. Furthermore, autonomous vehicles can communicate with each other and with traffic management systems, enabling smooth traffic flow and reducing the likelihood of accidents. This technology can also help reduce the number of vehicles on the road by enabling ridesharing and reducing the need for individual vehicle ownership.

DT technology can help urban and regional planners make better decisions by providing a virtual replica of the city or region. This DT can be used to simulate different scenarios and predict the outcome of different decisions. However, it is worth noting that the widespread adoption of autonomous vehicles would require significant investment and infrastructure changes, as well as addressing issues related to liability, privacy, and cybersecurity [22].

The use of autonomous vehicles augmented with AI can greatly reduce the number of accidents caused by human errors. The DT technology can also help in identifying potential risks in traffic and predicting possible accidents. With the help of sensors and cameras, autonomous vehicles can detect and avoid obstacles, other vehicles, and pedestrians, making them safer than traditional vehicles driven by humans. Technology has the potential to significantly reduce accidents on the roads. For instance, autonomous vehicles have the ability to detect and react to potential dangers faster and more accurately than human drivers. In addition to this, smart infrastructure can help to improve safety on the roads by providing real-time information on traffic conditions and potential hazards. DT technology can also be used to simulate and predict potential safety hazards, allowing for initiative-taking measures to be taken to prevent accidents [23].

Autonomous vehicles can reduce the risk of accidents caused by human errors such as drunk driving, distracted driving, or tiredness. They can be programmed to maintain a safe distance from other vehicles, follow traffic rules, and react quickly to changes in traffic conditions. Moreover, they are equipped with advanced sensors that allow them to "see" and "interpret" the environment around them, including road signs, other vehicles, and pedestrians. This enables them to make better decisions and avoid collisions even in low vision conditions.

DTs can help to simulate various real-world scenarios and test autonomous vehicles in a virtual environment. This will allow manufacturers to train the AI algorithms of the vehicles without the need for physical testing in different locations and under different scenarios. With DTs, manufacturers can also create realistic 3D models of the environment, which can be used to simulate and analyze various factors such as traffic density, pedestrian flow, weather conditions, and other variables that affect driving conditions. This can help in designing more effective autonomous systems and improve their safety in real-world conditions.

Improving public transportation is a key factor in reducing traffic congestion and improving mobility in cities. A well-designed and efficient public transportation system can encourage people to leave their cars at home and opt for more sustainable modes of transportation. This not only helps to reduce traffic on the roads, but also reduces air pollution and greenhouse gas emissions, and improves public health. Additionally, with the use of DT technology, public transportation systems can be optimized and made more efficient in terms of scheduling, route planning, and capacity management. The ways to increase the smartness and quality of services of smart cities were discussed. They discussed how the collaboration of using drones and IoT devices plays a great role in supporting smart cities. They also discussed how this DT technology helps in managing and efficiently using applications for communication, transportation, agriculture, waste management, disaster management, weather forecasting, etc.

The combination of blockchain, DTs, and AI has a significant impact on the management of urban cities. Blockchain technology can provide a secure and transparent system for storing and sharing data related to the city's infrastructure, while DTs can simulate and optimize the city's operations in real-time. AI can further enhance the system by analyzing the data and providing insights for better decision-making. The combination of these technologies can improve various aspects of urban management, including transportation, energy management, waste management, and more.

By utilizing the data obtained from the IoT technology and blockchain, the transport department can gain insights into the usage patterns of private cars and can offer discounts or incentives to encourage people to use public transport. This can help reduce traffic congestion and improve overall traffic management in urban areas. Additionally, it can also lead to a reduction in air pollution and carbon emissions, contributing to a cleaner and healthier environment.

With IoT, smart city operations can be simulated and with blockchain, it can be implemented securely [24]. With IoT, the city can be modeled virtually and many questions as discussed below can be answered through this virtual model:

- How to reduce congestion during rush hours?
- Can the introduction of more bike lanes reduce or increase traffic?
- What can be done to reduce pollution?
- How to handle waste management?
- How to supply necessary items to the public like water?
- How to plan flood management?

By connecting the DTs of physical objects in smart cities through the internet using blockchain, it is possible to exchange real-time data and improve mobility and governance. This can enable better management of resources and services, such as transportation, energy, waste management, and more. With real-time data and insights, decision-makers can make more informed decisions, leading to more efficient and effective management of the city.

1.2.3.2 MANAGING E-GOVERNANCE

Blockchain technology can help to eliminate intermediaries in digital transactions, making the process faster, more efficient, and more secure. This can reduce bureaucratic procedures and streamline interactions between government agencies and citizens. It can also reduce costs associated with intermediaries and intermediaries' errors, leading to greater transparency and trust in public services [25].

The combination of DTs and blockchain technology can bring transparency, security, and efficiency to urban city management services, including traffic management, public transportation, and governance. Blockchain technology can help facilitate secure and transparent data exchange between government and citizens, while DTs can provide an accurate and real-time simulation of the physical objects and infrastructure in a city.

This can help urban city development planners and authorities to make more informed decisions, optimize resources, and improve the quality of life for citizens [26].

1.2.3.3 MANAGING ELECTRIC SUPPLY

With the application of Blockchain and DT technology, households with solar panels can generate surplus electricity and share it with other households in the grid, creating a decentralized energy system. This can help reduce the overall demand for energy and promote sustainability. Blockchain technology can be used to track energy transactions between households, ensuring that the energy is distributed fairly and transparently. The DT technology can be used to simulate the energy demand and supply in the grid, helping utilities and households to better manage their energy usage and reduce waste [27].

1.2.3.4 MANAGING WASTE

The use of blockchain and DT technology can greatly improve various services in a city, including waste management, water supply, air quality control, and park and garden management. By utilizing sensors and data analysis, DTs can provide real-time information about the status of city infrastructure and services. Blockchain can ensure the security and privacy of this data and allow for more efficient and transparent interactions between city authorities and citizens. This can lead to more sustainable and effective urban planning and management.

Figure 1.3 shows the interconnection of IoT devices with AI and other technologies to achieve the best smart city. The physical layer of a smart city infrastructure typically consists of various physical objects like sensors, actuators, and other devices that communicate through IoT protocols such as Bluetooth, Wi-Fi, LAN, and WAN [28]. Blockchain technology can be implemented in the software layer to provide a secure and decentralized way of managing data and transactions related to smart city services. Different blockchain software platforms such as Ethereum, NXT, and Hyperledger Fabric can be used for this purpose. Depending on the needs of the specific application, the blockchain implementation can be either public or private. The ultimate goal is to create a smart city that is more efficient, sustainable, and responsive to the needs of its citizens. Traffic management systems like RITM2 (SIMETRA), Traffic maps, Intelligent Roadway Information Systems, KITS, etc., were examined.

A model of an efficient transport system was developed, which includes an information system for collecting and storing historical data of transport, road network, and AI algorithms for processing and decision-making. The visualization of transport activity, passenger flows, and mobile objects like pedestrians and vehicles can provide valuable information for urban and regional planners to optimize transportation systems and improve traffic management. This information can be used to better understand the transportation patterns of a city, identify areas of congestion, and implement solutions to reduce traffic and increase safety. In addition, communication and coordination with various emergency services such as the police and fire department are crucial for efficient and effective incident management in urban areas. The use of IoT technology and blockchain can help facilitate this communication and enable real-time tracking and monitoring of incidents, allowing for a quicker and more coordinated response [29].

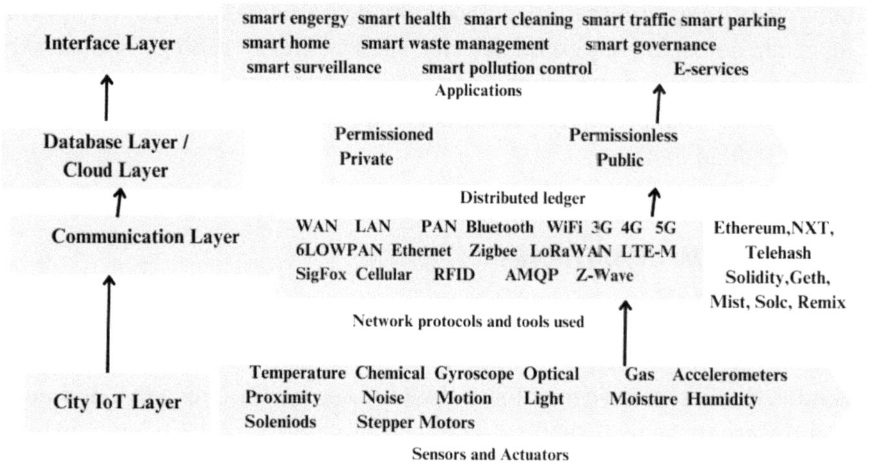

FIGURE 1.3 Smart city security framework.

1.2.4 *DTS IN MINING*

The use of DTs in remote monitoring and management of mine sites can help reduce operational costs and improve efficiency. By creating a digital replica of the physical mine site, operators can monitor and analyze real-time data from remote locations and use this information to identify potential issues, optimize processes, and make more informed decisions. This can also help improve safety by allowing operators to identify and respond to potential safety hazards in real time. The reliability of data generated on the mine site and immutably delivered to the DT is validated by blockchain technology.

DTs can be used to create a virtual replica of a mine or manufacturing site, which can be used for simulations, planning, and safety training. By creating virtual scenarios, workers can be trained in potentially dangerous situations without the risk of injury. In addition, DTs can be used to optimize production schedules and equipment maintenance, leading to increased productivity and reduced downtime. Overall, DTs have the potential to significantly improve safety, efficiency, and profitability in a variety of industries, including mining and manufacturing [30].

The use of DTs is significant in verifying the authenticity of commodities in supply chains. When coupled with Blockchain technology, the movement of a product can be monitored from its origin. DTs replicate the intended path of commodities, while Blockchain ensures that their behavior aligns with the DT, without any deviations. This tracking mechanism can detect any unlawful activity, such as illegal mining, by observing any inconsistencies in the behavior of the commodities. By implementing this system, manufacturers can track their projects globally and significantly reduce theft.

1.2.5 DTS IN HEALTHCARE

When a patient in critical condition seeks medical attention from a new doctor, the doctor often requests a series of tests to assess the patient's condition, even if these tests have already been performed and documented at a different medical facility. This can be due to a lack of connectivity and interoperability between different healthcare providers and information systems. As a result, the patient may have to undergo redundant tests, leading to unnecessary healthcare costs and a waste of time.

In healthcare systems, patient histories are digitally recorded as a chain of blocks in a network known as a distributed ledger. This ledger is secured by cryptography and timestamped to ensure its integrity and accuracy. The ledger is contributed to by various healthcare providers, such as nurses, surgeons, doctors, and lab technicians. Due to its decentralized nature, this data cannot be easily hacked or altered by any single party, since any changes made to the ledger must be validated by the other contributors in the network. This ensures that the data is reliable and secure and cannot be reversed or manipulated [31].

A simple use case of healthcare:

- **Step 1**: The information of the patient is collected by the healthcare providers.
- **Step 2:** The patient information is stored in the database.
- **Step 3:** A hash key is created for each block of data and is given to the blockchain.
- **Step 4:** The patient has key access to maintain the privacy of his/her data and he decides whom to give access.
- **Step 5:** Once a patient has permitted a doctor to access their medical data, the smart contract code present in each block of the distributed ledger is executed. The code verifies the permission and authenticates the doctor's identity. If the doctor's identity is authenticated and permission is granted, the hash key is matched with the specific block where the data is stored. The smart contract then allows the doctor to access the patient's data in that block, ensuring that the data is secure, tamper-proof, and only accessible to authorized parties.

Using blockchain technology in healthcare has two primary benefits. The first is that the system is patient-centric, which means that patients own their medical records. However, they cannot modify or delete any data within the blockchain. The data is created by healthcare providers, and patients have access to it. Furthermore, no one can access a patient's information without their explicit permission, which solves privacy concerns within the healthcare system.

The second advantage is that blockchain stores all the medical history of a patient in a secure, decentralized ledger. This means that when a patient goes to a new doctor or healthcare provider, they can access the patient's complete medical history on the blockchain, without having to repeat any tests or procedures that have already been done. This improves the efficiency and effectiveness of healthcare delivery, as doctors can provide better, and more informed care based on the patient's comprehensive medical history.

The outbreak of COVID-19 insists on using emerging technologies to handle the paramedic crisis. In order to acquire real-time data safely and securely, it is combined with blockchain technology, so when a case is referred worldwide the blockchain assures only the history of data is present everywhere and it is immutable. A prototype of blockchain-based collaborative DT framework was discussed which fights against COVID-19 and forthcoming diseases and describes a decentralized epidemic alert system [32].

In Ref. [33], the authors have suggested various protocols for implementing the Internet of Things in Medicine (IoMT). Discussions about how the medicine smart devices are interconnected for improving patient services were done and the application of various protocols like ZigBee protocol, Radio Frequency Identification (RFID), Time-Synchronized Mesh Protocol (TSMP), Near Field Communication (NFC), Wireless HART, etc., which are used to implement IoMT were explained.

Healthcare 4.0 utilizes DTs, blockchain technology, and the IoT to create an autonomous and data-driven healthcare system that enables decision-making by medical practitioners. In this system, data flows among interconnected devices in a decentralized and collaborative manner, allowing for a secure and timely exchange of information that can inform medical decision-making.

The predictive analytics that result from the amalgamation of these technologies can help medical practitioners make important decisions, such as recommending quarantine measures, lockdowns, and other restrictions in response to emerging health crises. This technology has the potential to significantly improve healthcare delivery by enabling more accurate and informed decision-making based on real-time data analysis.

1. **Interoperability:** In a decentralized system like healthcare 4.0, different models are defined in various places within the system. Data from both internal and external sources are received by these models. To ensure accurate decision-making, it is important to have a collective and standardized DT data schema that allows for effective communication between the different models.

   By using a standardized schema for DT data, different models within the decentralized system can work together in a coordinated and collaborative way. This ensures that all data is properly identified and accounted for, resulting in more accurate and informed decision-making across the entire healthcare system.

2. **Authentication:** In a decentralized environment where DTs are owned by independent entities, real-time and secure data exchange, and analysis is critical. To ensure that the system works properly, there needs to be a reliable and efficient mechanism for exchanging data in a secure and timely manner.

   This can be achieved through the use of advanced technologies like blockchain and the IoT, which allow for secure and decentralized data exchange. By leveraging these technologies, the system can ensure that all data is accurate and up-to-date and that it can be analyzed in real-time to inform decision-making and improve the overall efficiency of the healthcare system.

3. **Distributed Machine Learning:** In Healthcare 4.0, machine learning algorithms are utilized to analyze and predict the large amounts of data received from multiple DTs in the most effective way possible. Machine learning algorithms can help identify patterns and trends within the data that might be difficult or impossible for humans to detect. By analyzing the data in real-time, these algorithms can provide valuable insights and help healthcare providers make more informed decisions. This can lead to better health outcomes for patients and a more efficient healthcare system overall.

4. **Decentralized Decision-Making:** One of the key advantages of a decentralized healthcare system like healthcare 4.0 is that decisions can be made with the consensus of various independent entities within the system. Because the DTs are owned by independent entities, decisions must be made through a process of collaboration and consensus-building. This ensures that all stakeholders have a voice in the decision-making process and that decisions are made with a full understanding of all relevant data and perspectives. By leveraging the collective intelligence of all stakeholders within the system, decentralized healthcare can help to create a more efficient, effective, and equitable healthcare system for all.

5. **Scalability and Robustness:** To ensure that the Healthcare 4.0 system can handle various independent entities and their respective DTs, it must be designed to be robust and capable of producing a combined output.

Because the system must maintain data from multiple independent sources, it must be able to handle and integrate diverse types of data in a way that ensures the accuracy and integrity of the data. To achieve this, the system must be designed with robustness in mind, with built-in redundancy and backup systems to ensure that data is not lost or corrupted. By combining data from multiple sources, the system can provide a more complete and accurate picture of a patient's health, which can lead to better outcomes and a more efficient healthcare system overall.

Most of the above problems related to secure data can be solved by augmenting DTs with blockchain technology. A framework for the COVID-19 pandemic alerting system which involves the collaboration of blockchain, and DT technology was developed and discussed [34].

In a broader sense, the collaborative blockchain-based DTs to combat COVID-19 are depicted in Figure 1.4. The healthcare system is a combination of the physical and cyber worlds. The physical world includes doctors, patients, and other healthcare professionals, while the cyber world includes technologies like machine learning, CC, blockchain, DTs, AI, and big data. These two worlds work together to improve decision-making processes for hospitals, health organizations, and governments.

One technology that is particularly useful in the healthcare system is the Directed Acyclic Graph (DAG), which uses distributed ledger technology (DLT) to ensure that all participants have the same information. This leads to more accurate predictions that can be agreed upon worldwide. DAG also provides better control over medical service providers, patients, and practitioners.

In addition, supply chains in the pharmaceutical industry can be built with transparency using blockchain technology. Smart contracts can also be used to streamline the claims process for insurance companies, ensuring that patients receive their benefits quickly and efficiently.

Many researchers proposed a semantic and scope of healthcare DT and a patient-centric mathematics was developed using a cloud data model which implements blockchain technology. They described the system protocol, its analytic features, and its feasibility. Related data of a patient can be collected for the DT to analyze and provide exact reasons for the current state of the patient and also can predict the problems they may face in the future.

Public and private are two types of blockchain technology available. In a public blockchain, anyone can participate as a node, validate transactions, and write data onto the blockchain. This makes it a permissionless blockchain, meaning there are no restrictions on who can participate. Bitcoin and Ethereum are examples of popular public blockchain systems used for various applications.

In healthcare, there are also public blockchain systems that have been developed for storing and sharing medical data securely and transparently. Examples of public healthcare blockchains include MedRec, Nebula Genomics, FHIRChain, BurstIQ, and Guardtime, among others. These systems aim to improve data sharing and interoperability across various healthcare providers and institutions while maintaining patient privacy and security.

Private blockchains are usually used in healthcare systems because of their enhanced security and privacy features. These systems require strict access control and privacy measures to protect sensitive health data, and private blockchains can provide these capabilities. They can also facilitate the interoperability of health data between different entities securely and transparently. Some prominent private blockchains are Hyperledger Fabric, Hyperledger Sawtooth, Corda, etc. Prominent blockchain-based healthcare systems are Health Chain, Model Chain, Ancile, MeDShare, etc.

A framework that is a collaboration of blockchain-based DT for providing an alerting service for COVID-19-like pandemics such as the declaration of quarantine, lockdown protocols, etc., was developed by authors [35]. They defined their framework with four layers, namely in the high level – physical layer, next level – blockchain-based DTs layer, intermediate level – data analytics layer, and at the core level – decision-making layer. The physical layer would likely involve humans (doctors, nurses, pharmacy, government sector, etc.) and non-human participants (IoT devices like CCTV, sensors for measuring blood pressure, ventilator, ECG, etc.) that collect patient data, which would then be fed into the DT layer built on top of blockchain technology. The data analytics layer would use machine learning and other algorithms to process and make sense of the data, while the decision-making layer would use this information to make clinical decisions and recommendations. Each layer would likely have its unique challenges and opportunities for innovation as exposed in Figure 1.4.

Next is the blockchain-based DT which consists of DTs connected with blocks in the network. This node information for DTs is supplied by human and non-human participants. The data is registered by DT in terms of timestamp, owner of creation of information, status (sensor readings), and transaction as a digital ledger. These IoT devices with sensors generate lots of digital data which they communicate with each other securely through this blockchain network. The layer in the next level is the data analytics layer which consists of machine learning algorithms that receive data from the distributed ledger and give accurate predictions for decision-making. The data analytics is done with the historical data stored with timestamps and live streaming data received from DT.

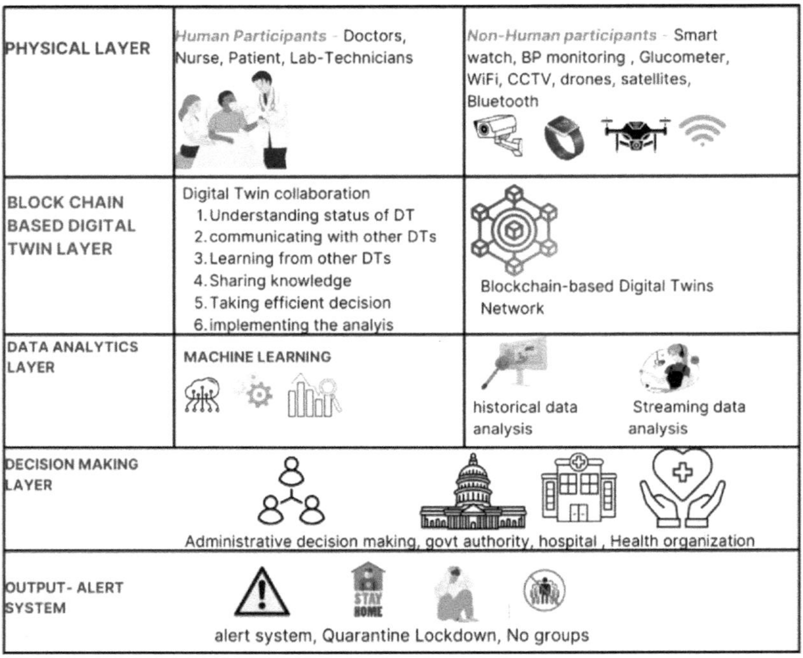

FIGURE 1.4 Architecture model of block-chain based DT for healthcare management.

The next layer is the decision-making layer where the collaboration with various agencies to give various timely outputs like pandemic outbreaks, information for announcing quarantine, etc. It is not practical to store all patient data in every block of the blockchain. Data lakes are commonly used to store large amounts of unstructured and semi-structured data. They are designed to provide scalable storage that can handle both structured and unstructured data and allow for easy retrieval and analysis of the data. In a healthcare context, data lakes can be used to store patient data that is not needed for every transaction on the blockchain but may be needed for future analysis or decision-making. The blockchain can then be used to store key metadata about the patient data, such as timestamps and hashes, which can be used to ensure the integrity and authenticity of the data.

Data lakes can also be used to store data from other sources, such as medical devices, wearables, and other IoT devices, which can be used to provide additional context and insights into a patient's health status as demonstrated in Figure 1.5. The data stored in the data lake can then be analyzed using machine learning and other data analytics tools to identify patterns and trends that can be used to improve patient care and outcomes. The digital signature is a cryptographic technique used to verify the authenticity and integrity of a digital document, message, or transaction.

In the healthcare blockchain, when a medical examiner creates a medical record, a digital signature is created using a private key that is unique to the examiner. This digital signature ensures that the document is authentic and has not been tampered with. Other participants in the blockchain can then verify the digital signature using the examiner's public key to ensure that the document is valid and can be trusted. The patient can have access to their medical records and can contribute to them by adding information obtained through their personal health devices. The data stored in the data lake can be analyzed for insights, which can be used for personalized healthcare services and better decision-making. Access to this data is limited to authorized individuals and the blockchain ensures the integrity and security of the data.

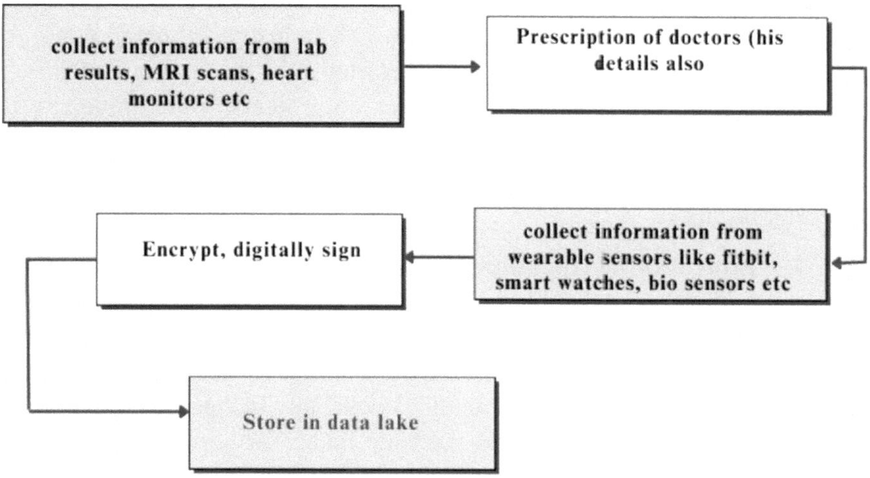

FIGURE 1.5 Storing of data in healthcare blockchain.

1.2.6 BLOCKCHAIN IN EDUCATION

Blockchain technology can be used to securely store and share educational credentials, as well as to provide a transparent and tamper-proof way of verifying the authenticity of educational and financial records. By storing student records on a blockchain, educational institutions can provide a secure and verifiable way for employers and other organizations to verify the credentials of students. This can greatly reduce the time and cost associated with background checks and other verification processes [36].

In addition, blockchain can be used to streamline the process of disbursing grants and scholarships. By using smart contracts, funds can be automatically disbursed to eligible students based on predetermined criteria, such as academic performance or financial need. This can help reduce the administrative burden associated with managing grant and scholarship programs, while also ensuring that funds are distributed fairly and transparently.

Blockchain technology enables the creation of a tamper-proof digital record that can be accessed by students and institutions alike. This eliminates the need for paper-based records that can easily get lost, damaged, or altered. By storing education records on a blockchain, students can have greater control over their data and share it with institutions or potential employers as needed. Educational institutions can also benefit from improved record keeping, as they can easily track student progress and ensure that their records are up-to-date and accurate.

In 1997, Peter Drucker, who was an Austrian–American management consultant, author, and educator, predicted that universities wouldn't exist in the future. Now, this is coming true with Industry 4.0 technology. An innovative coding university in Paris started in 2013 is proof of this. It has no lecturer, tutor, or books. Students do projects in the university by undergoing internship programs designed, and after completion of the projects, they get their degree by reaching a different level of learning by earning points.

1. **Education Verification:** Education institutions have to coordinate with other stakeholders' accreditation authorities to validate the academic records of applicants claiming to be successful students. This whole process of validating candidate credentials is time-consuming and costs money. Blockchain Technology can be used to reduce overhead and the time taken for certificate validation from days to seconds.

2. **Digital Degrees and Certificates:** Fraudulent degrees are an increasing problem around the world. People get a fake certificate/diploma in a few minutes for a low cost. Blockchain technology can be used to issue unique digital access that verifies the credentials of academic degrees and certificaticns. This would help the employer to check the originality of the certificates in less time.

   The prevalence of fraudulent degrees is on the rise worldwide, with individuals easily obtaining fake certificates and diplomas at a low cost. However, the implementation of blockchain technology could be a potential solution to combat this issue. Creating unique digital access that verifies academic credentials, degrees, and certifications can be authenticated much more efficiently. This would greatly benefit employers by providing a quicker and more accurate way of verifying the authenticity of certificates.
3. **Cloud Storage:** Online collaboration has become essential in the educational sector, leading educational institutions to provide vast amounts of data to students and faculty on cloud platforms such as Microsoft Azure, AWS Cloud, IBM Cloud, Alibaba, and Google Cloud Platform. However, these cloud platforms can be costly, and the expenses incurred are passed on to students in the form of fees. A viable alternative to cloud storage is blockchain-based file storage, which saves data in a decentralized network. By utilizing the unused disk space of users worldwide, this method can significantly reduce the cost of purchasing cloud storage.
4. **Access to Resources:** Blockchain technology enables the secure sharing of educational resources such as books, learning materials, videos, and podcasts on a public network for students to access and use.

1.2.7 BLOCKCHAIN IN CRYPTOCURRENCY

A cryptocurrency, or crypto, is a decentralized digital currency that uses blockchain technology to record and secure every single transaction. The "crypto" in cryptocurrency refers to cryptography that allows for the creation and processing of digital currencies that do not allow counterfeiting or spending multiple times [37].

The first decentralized cryptocurrency using blockchain technology created is Bitcoin (BTC). There are plenty of other cryptocurrencies known as altcoins (Alternative Coins) other than Bitcoin. Some of the altcoins are Tether (USDT), Ethereum (ETH), Cardano (ADA), USD Coin (USDC), Binance Coin, XRP, Litecoin, Polkadot, IOTA, Ripple, etc. During the year 2022, CoinMarketCap provided a source that there are around 22,000 cryptocurrencies worth $850 billion.

Few advantages of cryptocurrency:

- A cryptocurrency transaction is a transfer of a variety of cryptocurrencies anywhere in the world with little time on the blockchain.
- Transactions can be done easily through mobile, desktop, or hardware wallets at a low cost without having a bank account.
- Decentralized cryptocurrencies make secure systems as they use the techniques of cryptography and blockchain security.
- Crypto markets work round the clock without any break.

Disadvantages of cryptocurrency:

- Understanding cryptocurrency takes time and effort.
- It is difficult for the government to track the exchange of money involving illegal transactions.
- Crypto mining requires special hardware like Graphics Processing Units, Solid-State Drives for crypto mining, Application-Specific Integrated Circuits, mining software, and a wallet.
- Crypto mining cannot be done on regular computers.
- There is no guarantee of protecting our business as there is no involvement of the Financial Conduct Authority.
- There is a chance of stopping/moving to new cryptocurrency stakeholders.
- No reverse transaction can be created if someone mistakenly sends funds to the wrong wallet address.
- Cryptocurrencies are not accepted everywhere and have limited value.

1.2.8 BLOCKCHAIN APPLICATION IN VOTING

The foundation of any democratic country is its election process, through which citizens exercise their right to vote for the candidates they believe in. In India, representatives are elected through elections at all levels, from local bodies and panchayats to parliament. It is crucial that the election process is both transparent and reliable to maintain the credibility of the entire system.

If the process is only reliable, citizens will have confidence in the system and participate in voting, but if transparency is lacking, citizens may choose to ignore the system, which poses a significant risk to democracy [38].

In India, before 1997, the ballot box method was the primary means of conducting elections. Voters would mark their desired candidate's symbol on a printed electoral chit and cast their vote. However, in 1997, the electronic voting machine was introduced as an alternative method for conducting elections, and it has been the sole means of voting since 2004. The Election Commission of India is implementing continuous evaluation in the election process to ensure more efficient, transparent, and reliable elections. Despite the use of the Electronic Voting Machine system, fraudulent voting continues to occur in elections. To address these and other issues, the introduction of blockchain and DLTs in the electoral process could be beneficial. Challenges such as voter anonymity, vote confidentiality, and end-to-end verification for voters to confirm that the vote they cast is the one they intended can be effectively addressed and implemented with these technologies.

An e-voting scheme with a multi-chain platform implementation was proposed. Proponents of blockchain implementation in e-voting models argue that the cryptographic properties of the network can maintain the secrecy of individual voters. Through the use of a hashing technique to generate unique identifiers for each user and register them on the blockchain, double voting can be prevented. A model that ensures convenience through a web-based interface that requires minimum input from the user was proposed [39]. For example, instead of username/password, fingerprints are used so that even if they forget the username/password, it will not affect their login. A detailed framework is illustrated in Figure 1.6.

Horizon State, a company specializing in blockchain technology, developed e-voting software that has been used in elections. In November 2018, the Thailand Democrat Party became the first political party to use e-voting software based on blockchain technology, specifically the Zcoin platform. The Chief Election Commissioner of India, Mr. Sunil Arora, recently announced that remote voting will be introduced for the country's 2024 general election. To implement this technology, the Election Commission is collaborating with IIT-Madras and exploring the use of blockchain and DTs.

The Election Commission has suggested linking voter IDs with Aadhaar IDs and has sent the proposal to the Ministry of Law. Additionally, the Commission has recommended the incorporation of IoT devices with GPS chips into the existing EVMs to track their location, thereby preventing

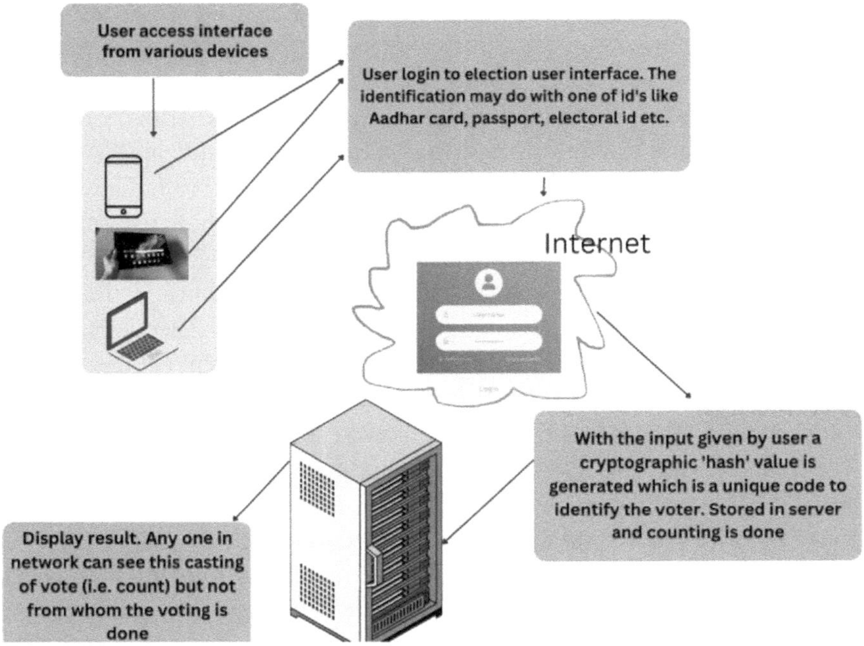

FIGURE 1.6 Blockchain based DT for e-voting.

tampering and theft of Electronic Voting Machines. The Election Commission of India has decided to offer digital election photo ID cards to voters, to improve voting percentage and provide faster delivery of electoral services.

1.3 BLOCKCHAIN AND DT HELPS IN SOLVING PROBLEMS IN INDIA

Corruption is the biggest threat and problem in India, and to address this issue, adopting blockchain-based DLTs could provide a transparent and tamper-proof system for tracking government spending on various schemes. Adopting blockchain-based DLT technology can provide solutions to various issues in India, including tracking of real estate lands, registration, identification of crimes, regularizing healthcare activities, disease control, supply chain management, traffic regulation, and implementation of smart cities.

How many central government schemes are adopting blockchain technology with DLT? A few are given below:

1. **Rashtriya Madhyamik Shiksha Abhiyan (RMSA):** In this, student enrollment details, school details, and students' mark sheet details are recorded digitally. When the students go for higher education, verification of their data has become easy. The distribution of many scholarships also becomes easy as all the details like Aadhar, mobile, and account details are stored digitally, and the scholarship money reaches the student directly to their account without any delay.
2. **Pradhan Mantri Kisan Sampada Yojana (PMKSY):** This provides a digital infrastructure with an efficient supply chain for farmers from sowing to retail outlets.
3. **Pradhan Mantri Kisan Samman Nidhi (PM-KISAN):** Through this scheme, an income of Rs. 6,000 per year in three equal installments is distributed to all the farmers into their bank accounts.
4. **Pradhan Mantri Fasal Bima Yojana:** In this scheme, farmers can get insurance for their crops, report crop loss, submit their grievances, and track the status of their complaint.
5. **The Rajasthan Government:** It was the first state to adopt day-to-day governance and one of the famous schemes in the adoption of blockchain technology with DT for maintaining electronic health records, named Bhamashah.
6. **The Ministry of Electronics and Information Technology (MeitY):** It has set up a Center of Excellence in blockchain technology. They, along with the National Informatics Center (NIC), work to develop many successful blockchain networks for various government schemes. The four successful products developed by them are Certificate Chain, Document Chain, Property Chain, and Logistic Chain. The government, in collaboration with the Netherlands, formed a blockchain application incubator called Quantoz to provide secure digital payments.
7. **The Maharashtra Government:** It has revamped documentation by leveraging distributed ledgers for land properties.

8. **Punjab:** It is using blockchain in farming to track seeds of potatoes from nucleus to harvest.
9. **Uttar Pradesh:** In Uttar Pradesh, the Firozabad Police adopted the Polygon blockchain to register FIRs
10. **StaTwig Company:** It implemented blockchain to track each vial using a vaccine ledger in the network
11. **Andhra Pradesh:** It has collaborated with a Swedish company and developed a blockchain-based platform to maintain land registration digitally. This enables farmers as well as the public to be aware of land fraud.
12. **Start-Up India:** It has launched a blockchain-based certificate verification system for maintaining the authenticity of certificates awarded to start-ups. Private sector organizations, angel investors, and Venture Capital System use this method to verify their certificates and enable funding for start-ups in a swift and tamper-proof way.
13. **The Kerala Blockchain Academy (KBA):** It is working on using blockchain technology for storing and verifying educational certificates. By creating a DT of the certificate on a blockchain network, the system ensures that the certificate is tamper-proof and can be easily verified. The use of QR codes, timestamps, and block numbers ensures that the authenticity of the certificate can be quickly confirmed using a mobile phone. This will provide a global verification mechanism for educational institutions and employers.
14. **As a Trial, the Telangana State Election Commission (TSEC):** It decided to apply a face recognition system in a few election booths to avoid the offense of undue influence or impersonation during elections.

1.4 CONCLUSION

Blockchain and DTs have the potential to revolutionize many fields, including logistics, manufacturing, healthcare, and more. A DT replicates the processes of any real-life objects like buildings, cities, machines, windmills, etc., and simulates them to help gather the history of their activities and predict their performance. DTs can help organizations monitor and optimize the performance of physical assets, while blockchain can provide secure and transparent record-keeping for transactions and data sharing. The records on a blockchain are secured from hackers through cryptography. The amalgamation of blockchain and DTs leads to strong, credible, efficient, and correct decision-making in many real-time applications like logistics, where the utility industry leverages for improved productivity. Blockchain technology monitors the safety of the processes created by DTs.

KEYWORDS

- **artificial intelligence**
- **blockchain**
- **digital twin**
- **internet of things**
- **smart contract**
- **virtual reality**

REFERENCES

1. Cannavacciuolo, L., Ferraro, G., Ponsiglione, C., & Primario, S. (2023). Technological innovation-enabling industry 4.0 paradigm: A systematic literature review. Technovation, 124, 1–16.
2. Saurabh, V., Prashant, A., & Santosh, B. (2018). Industry 4.0–A glimpse. Procedia Manufacturing, 20, 233–238.
3. Vazhuthi, P., Aruchamy, P., & Manikandan, S. P. (2022). An energy-efficient auto clustering framework for enlarging quality of service in internet of things-enabled wireless sensor networks using fuzzy logic system. Concurrency and Computation: Practice and Experience, 34, 1–26.
4. Dhiviya, S., Malathy, S., & Kumar, D. R. (2018). Internet of things (IoT) elements, trends and applications. Journal of Computational and Theoretical Nanoscience, 15, 1639–1643.
5. Muneeswari, G., Sajithra, S., & Hegde, R. (2023). Self-diagnosis platform via IoT-based privacy preserving medical data. Measurement: Sensors, 25, 1–10.

6. Tissir, S., Cherrafi, A., & Chiarini, A. (2023). Lean six sigma and industry 4.0 combination: Scoping review and perspectives. Total Quality Management & Business Excellence, 34, 261–290.
7. Mathur, A., Kalla, A., Gur, G., & Bohra, M. K. (2023). A survey on role of blockchain for IoT: Applications and technical aspects. Computer Networks, 227, 1–12.
8. Sasikumar, A., Vairavasundaram, S., Kotecha, K., & Indragandhi, V. (2023). Blockchain-based trust mechanism for digital twin empowered Industrial Internet of Things. Future Generation Computer Systems, 141, 16–27.
9. Bhaskar, K. B., Prasanth, A., & Saranya, P. (2022). An energy-efficient blockchain approach for secure communication in IoT-enabled electric vehicles. International Journal of Communication Systems, 35, 1–20.
10. Liu, S., Lu, Y., Li, J., & Shen, X. (2023). A blockchain-based interactive approach between Digital Twin-based manufacturing systems. Computers & Industrial Engineering, 175, 1–16.
11. Mathew, T., Sabu, A., & Sengan, S. (2023). Microclimate monitoring system for irrigation water optimization using IoT. Measurement: Sensors, 27, 1–12.
12. Qimeng, L., Gravina, R., Ye, L., & Saeed, H. (2020). Multi-user activity recognition: Challenges and opportunities. Information Fusion, 63, 121–135.
13. Zhao, R., Chen, Z., & Xue, F. (2023). A blockchain 3.0 paradigm for digital Twins in construction project management. Automation in Construction, 145, 1–10.
14. Schluse, M., Priggemeyer, M., Atorf, L., & Rossmann, J. (2018). Experimentable digital twins—Streamlining simulation-based systems engineering for industry 4.0. IEEE Transactions on Industrial Informatics, 14, 1722–1731.
15. Sadri, H., Yitmen, I., Tagliabue, L. C., & Westphal, F. (2023). Integration of blockchain and DTs in the Smart Built Environment Adopting Disruptive Technologies—A Systematic Review. Sustainability, 15, 1–10.
16. Lo, C. K., Chen, C. H., & Ray, Y. Z. (2021). A review of digital twin in product design and development. Advanced Engineering Informatics, 48, 1–10.
17. Broo, D. G., & Schooling, J. (2021). Digital twins in infrastructure: Definitions, current practices, challenges and strategies. International Journal of Construction Management, 23, 1254–1263.
18. Capocasale, V., Danilo, G. B., & Perboli, G. (2023). Comparative analysis of permissioned blockchain frameworks for industrial applications. Blockchain: Research and Applications, 4, 1–10.
19. Liu, M., Fang, S., Dong, H., & Xu, C. (2021). Review of digital twin about concepts, technologies, and industrial applications. Journal of Manufacturing Systems, 58, 346–361.
20. Rawat, B., Bist, A. S., Apriani, D., & Permadi, N. I. (2022). AI based drones for security concerns in smart cities. APTISI Transactions on Management (ATM), 7(2), 125–130.
21. Wang, H., Chen, X., Jia, F., & Cheng, X. (2023). Digital twin-supported smart city: Status, challenges and future research directions. Expert Systems with Applications, 217, 1–12.
22. Karale, S., & Ranaware, V. (2019). Applications of blockchain technology in smart city development: A research. International Journal of Innovative Technology and Exploring Engineering (IJITEE), 8, 1–12.

23. Ibrahim, M., Rjabtšikov, V., & Gilbert, R. (2023). Overview of digital twin platforms for EV applications. Sensors, 23, 1–14.
24. Jafari, M., Kavousi-Fard, A., Chen, T., & Karimi, M (2023). A review on digital twin technology in smart grid, transportation system and smart city: Challenges and future. IEEE Access, 11, 17471–17484.
25. Li, S. (2018). Application of blockchain technology in smart city infrastructure. IEEE International Conference on Smart Internet of Things (SmartIoT), 1, 276–2766.
26. Kandpal, V., Chandra, D., Dalei, N. N., & Handoo, J. (2023). Expanding financial inclusion through fintech and e-governance. In Financial Inclusion in Circular Economy. Circular Economy and Sustainability. Springer Cham, 1, 1–20.
27. Mhamud, H. S., Safwat, M. C., & Sajal, K. D. (2023). Towards electric digital twin grid: Technology and framework review. Energy and AI, 11, 1–15.
28. Shantha, R., Mahender, K., & Jenifer, A. (2022). Security analysis of hybrid one time password generation algorithm for IoT data. AIP Conference Proceedings, 2418, 1–10.
29. Vorobeva, D., Scott, I. J., & Oliveira, T. (2023). Leveraging technology for waste sustainability: Understanding the adoption of a new waste management system. Sustain Environ Res, 33, 1–12.
30. Wang, J., Huang, Y., Zhai, W., & Li, J. (2023). Research on coal mine safety management based on Digital Twin. Heliyon, 9, 1–10.
31. Akash, S. S., & Ferdous, M. S. (2022). A blockchain based system for healthcare digital twin. IEEE Access, 10, 50523–50547.
32. Sahal, R., Alsamhi, S. H., Brown, K. N., & O'Shea, D. (2022). Blockchain-based digital twins' collaboration for smart pandemic alerting: Decentralized COVID-19 pandemic alerting use case. Computational Intelligence and Neuroscience, 10, 1–14.
33. Prasanth, A., Lakshmi, D., Rajesh, D., & Balamurugan, B. (2022). Cognitive computing for internet of medical things. Chapman and Hall/CRC, 1–230.
34. Maged, N., Boulos, K., & Zhang, P. (2021). Digital twins: From personalized medicine to precision public health. Journal of Personalized Medicine, 11, 1–12.
35. Moztarzadeh, O., Jamshidi, M., Sargolzaei, S., & Keikhaee, F. (2023). Metaverse and medical diagnosis: A blockchain-based digital twinning approach based on mobilenetv2 algorithm for cervical vertebral maturation. Diagnostics, 13, 1–12.
36. Mohammad, A., & Vargas, S. (2022). Challenges of using blockchain in the education sector: A literature review. Applied Sciences, 12, 1–20.
37. Cui, W., & Gao, C. (2023). WTEYE: On-chain wash trade detection and quantification for ERC20 cryptocurrencies. Blockchain: Research and Applications, 4, 1–15.
38. Jafar, U., Aziz, M. J. A., & Shukur, Z. (2021). Blockchain for electronic voting system—Review and open research challenges. Sensors, 21, 1–10.
39. Khan, K. M., Arshad, J., & Khan, M. M. (2018). Secure digital voting system based on blockchain technology. International Journal of Electronic Government Research (IJEGR), 14, 53–62.

CHAPTER 2

Research Challenges and Opportunities in Blockchain Technology

B. RAGAVI,[1] S. SHARMILA,[1] A. MOHAN KUMAR,[2] and SEIFEDINE KADRY[3–5]

[1]*Department of Biomedical Engineering, Dr. N.G.P. Institute of Technology, Coimbatore, Tamil Nadu, India*

[2]*Tata Consultancy Private Limited, Chennai, Tamil Nadu, India*

[3]*Department of Applied Data Science, Noroff University College, Kristiansand, Norway*

[4]*Artificial Intelligence Research Center (AIRC), Ajman University, Ajman, United Arab Emirates*

[5]*Department of Electrical and Computer Engineering, Lebanese American University, Byblos, Lebanon*

ABSTRACT

Artificial Intelligence (AI) in Blockchain technology is one of the most highlighted advancements in recent days, and it encourages AI toward intelligence as more enhancement to improve the flexibility and reliability of the distributed ledger. It is a rapidly gaining stream that works under versatile technology and is widely used in various fields with intelligent decision-making capability for a machine. The data from multiple repository platforms has been collected, stored, and transmitted in a secure and trust-level manner. In contrast, it provides a distributed decentralized storage by distributing the records, data updates, and storage that might be performed without a single centralized point. It can

Blockchain-Based Digital Twins: Research Trends and Challenges.
A. Prasanth, Vandana Sharma, Rajesh Kumar Dhanaraj, Pethuru Raj (Eds.)

automate the healthcare industry, such as electronic health records, pharmaceutical industry, cyber security, tracing, and tracking platforms for Supply chains for smart contracts and ledger data. These data are interrelated with sensors and other network devices facilitating development and productivity. Further, Blockchain-based AI algorithms rely on data to learn, infer, and make efficient decisions for future advancements in the research area. These algorithm implementations improved efficiency for various applications, and it is insufficient to enhance the development of QoS such as Scalability, persistence, anonymity, and privacy. To overcome these limitations, this chapter gives a detailed summary of open research challenges for different applications. It ensures the possibilities of security enhancement with outstanding learning performance and confidential information. The insightful explanation emphasizes blockchain-based frameworks in recent research and challenges, and it will highlight the pros and cons of different concepts. Furthermore, this chapter deals with open research challenges under various AI applications and Blockchain technology.

2.1 INTRODUCTION

The most prominent development in current history is blockchain, which has been extensively used in most sectors. In 2008, it emerged as a disruptive innovation along with a transparent mechanism. In contrast, it is used to transact bitcoin and cryptocurrency on a ledger and make transactions electronically with a controlled mechanism. At the same time, due to the absence of centralization, it is more concerned with developing and maintaining traceability, security, and confidentiality. Recently, IBM has proven that blockchain technology in the healthcare sector and other related areas promotes better medical service. It is expected to earn more than 500 million USD by 2022 and has a greater potential for developing healthcare information frameworks. It has tremendously increased the healthcare and IT sectors with several concepts such as decentralization, peer-to-peer networks, Merkle, mining, and consensus algorithms [1].

In blockchain technology, the decentralized token system provides authentication with high computational power. Generally, it has two types of data chains: on-chain and off-chain data. On-chain data stores all the data in the block, while off-chain data cannot store the data in the block. Compared to on-chain, off-chain has less complexity in transacting information and maintaining transactional records in any configuration. In contrast, it cryptographically stores subsets of chaining records in the form of data blocks to their temporal predecessors. Blockchain provides robust security and has a high potential for distributing data [2, 3]. Some cryptographic algorithms are used to enhance the security level for peer-to-peer users. Despite blockchain's enormous potential, there are a few complications and limitations, such as single-point failure, high transaction fees, malicious attacks, low latency, and trust issues. This chapter mainly aims to highlight open research challenges in blockchain technology by describing concepts with various applications.

2.1.1 CHARACTERISTICS AND LIMITATIONS IN BLOCKCHAIN TECHNOLOGY DECENTRALIZATION

Decentralization in Blockchain technology is used in each node, providing reliability and recognized authorities in unreliable circumstances. In decentralization, the distributed network controls the central entities, lowers the level of trust among participants, hinders the capacity to control the external authority, and deteriorates the network's functionality shown in Figure 2.1.

While lowering the level of trust, data reconciliation occurs in the system, and it reduces the point of weakness which leads to systematic failure, such as bottlenecks with inadequate incentives for trusting the nodes [4]. For example,

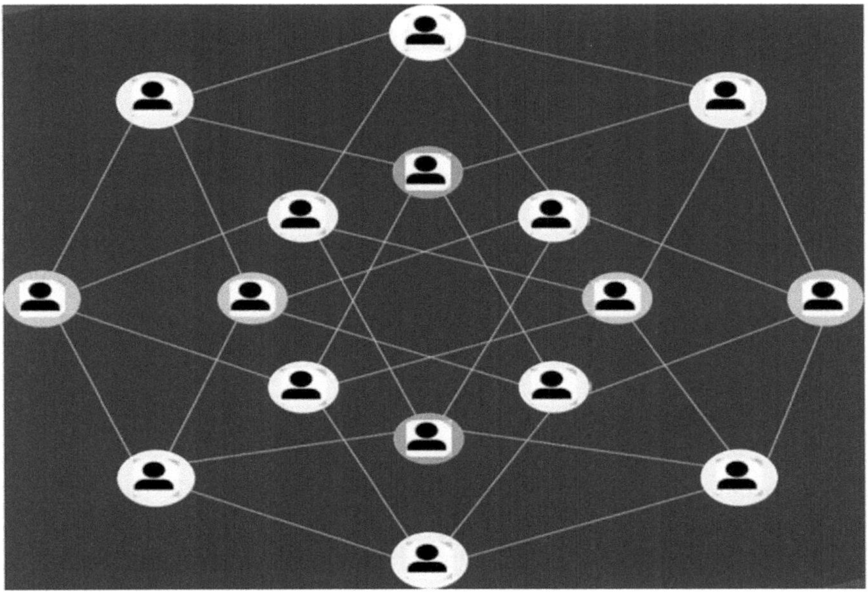

FIGURE 2.1 Decentralization of blockchain technology.

banking systems follow centralized trusted authorities because it has minimal transaction costs. Moreover, decentralization has a few drawbacks, such as decreased transaction throughput and high latency. Still, in an ideal world, these drawbacks are justified by the gain of instability and service level with common centralized and distributed networks.

2.1.2 PERSISTENCY

To provide temporal decentralization, the blockchain in a network's transactions is automatically stored in blocks with a distributed chain to the entire network, leading to latency and redundancy. Furthermore, by storing the data in many nodes, improper identification, falsifications, or missing data frequently occur during transactions. For example, digitally signed information integrity can only be confirmed if the relevant certificates still exist. However, the short lifespan of the certificates undermines their long-term resilience and verifiability.

2.1.3 ANONYMITY

In blockchain technology, banking systems always maintain and identify valid transactions along with the valid person's transaction details [5]. It can be easily tracked, but it does not provide enough anonymity for transactions. By using multiple addresses and interacting with the generated wallet address, a person can keep their identity private, and there is no central repository for a node's private data. In each transaction, the addresses are associated with wallet addresses, and the transactions are publicly verified while maintaining their privacy. Still, the transaction cannot assure who issued or owned a wallet's blockchain address. A central party does not request a user's personally identifiable data anymore.

2.1.4 TRANSPARENCY

By providing decentralized and consistent transaction records, the blockchain network ensures wider network transparency. The importance of transparency cannot be underestimated in the supply chain and logistics sector networks. By constructing a transparent system, the blockchain network reduces the number of forgeries, and it is obvious how the data are coordinated or automated in the procedures from inputs to several parties in smart contracts. Furthermore, the inherent transparency of blockchains frequently conflicts with organizational confidentiality rules and data protection regulations, which minimizes the transaction process.

2.1.5 LOSS OF PRIVACY

A significant amount of privacy is protected in blockchain using public key cryptography. For each transaction, authentication is maintained to protect the user identification. In contrast, the ledger collects the data and enables transactional security publicly for providing transactional anonymity because all transaction identifiers and balances for each cryptographic key are made available to the public. Still, data privacy works against privacy policies and procedures. Therefore, blockchain is more suitable for various fields internationally, even though it has not yet achieved its maximum development for improving privacy.

2.2 RESEARCH TRENDS AND OPEN CHALLENGES IN BLOCKCHAIN TECHNOLOGY

2.2.1 ELECTRONIC HEALTH RECORDS

Electronic Health Records digitally store health information, a revolutionary invention in the 20th century. In networks and communication, it offers distributed and decentralized networks among all the nodes without central authority by improving health record management. It provides privacy, security, and accessibility to distributed P2P networks and does not depend on a third entity. In distributed P2P networks, the digital data is allocated to all the users securely, publicly, or privately.

In conventional methods, both the sender and receiver acknowledge the transaction details depending on the trusted third party, which leads to few security trust issues and operational privacy with the lengthy transaction. But millions of data in blockchain are highly beneficial for healthcare providers, but the open research challenge for researchers is to fix the data breaches and also external attackers creating security issues while transmitting the information. Therefore, monitoring and providing suggestions for improving verification and validation is essential.

If the monitoring and suggestions are to be validated, the decentralized system for tracking data can be accessed safely when needed to avoid security threats. However, if the validation is not scrutinized properly, the authentication protocol cannot transmit the data. If the nonce value exceeds the intended value, the stored block is not eligible to be added to the blockchain. This addresses the limitation in EHR management, and it is essential to provide a secure and decentralized environment for exchanging data with the help of Artificial Intelligence and edge computing.

2.2.2 EHR WITH ARTIFICIAL INTELLIGENCE

Integrating blockchain with artificial intelligence techniques like machine learning and deep learning is more effective in healthcare applications. In contrast, it helps to automate and design real-time applications with greater potential to store and share data confidentially. Researchers are mainly focused on developing the machine learning technique in blockchain-based EHR applications. Recently, IBM has declared a few plans to implement blockchain with AI, where an AI agent performs multiple tasks, including law enforcement, updated recordkeeping, and illegal behavior, and encouraging improvements to smart contracts over a wide network [6]. A machine learning algorithm is used to record and identify fraudulent data to achieve the above parameters. With the help of the development of this algorithm, old corrupted medical records, drug analysis, and disease prediction can be retrieved and preserved.

2.2.3 EHR WITH EDGE COMPUTING

Edge computing is also the best choice to share and store enormous healthcare data based on network load and data size. It includes massive data storage, broad networking, and high computing power, and also supports distributed applications with flexibility in a controlled and secure approach [7, 8]. Recently, researchers have mainly focused on extending a cloud service to the network edge and providing high processing capability to optimize HER devices, as shown in Figure 2.2. Even though researchers are working on edge computing, it is a more essential issue to fix the storage, scalability, block size, and block creation time of various attacks while transacting data, security, integrity, and vulnerability.

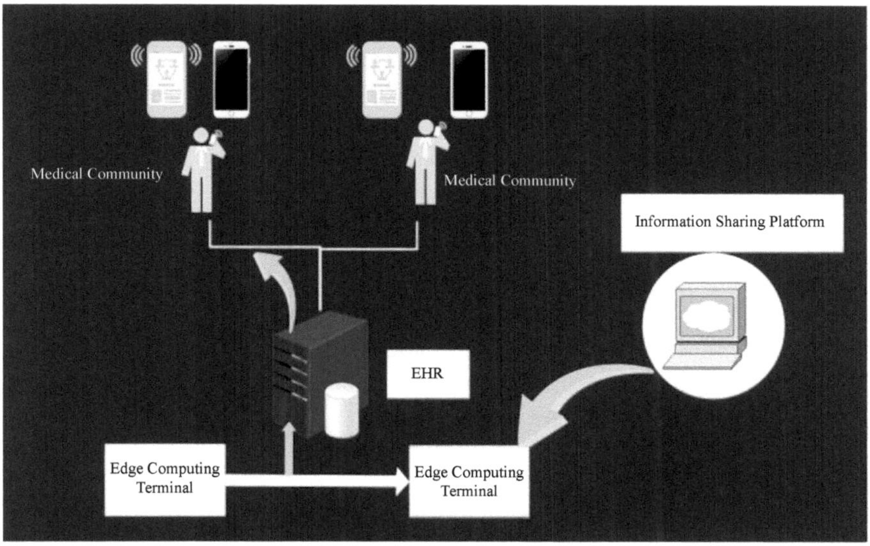

FIGURE 2.2 EHR information sharing to edge computing.

2.3 BLOCKCHAIN IN PHARMACEUTICAL INDUSTRY

Blockchain technology is an advanced technology in recent digital platforms. It improves different automated industries in such a way that the pharmaceutical industry suffers from a lack of simplicity and transparency, leading to transmission difficulties, including products that may reach an expired condition. Blockchain technology has adopted different applications like preventing forgery in drug distribution, enhancing security, monitoring transportation, ensuring data quality, governance, and turnover, and improving the perception of the drug with the distributed ledger. It makes it possible to record transactions among various nodes in a trustworthy, sustainable, and efficient manner. Also, security systems such as encryption and decryption under cryptography are used to ensure data accuracy.

Moving to the next level of the system, Blockchain technology includes authentication with verifiable serial codes for the source document. With the help of the serial code, drug quality control, and tracking are carried out with digital signatures deployed from the producer to the pharmacy to maintain the stable quality of the drug and prevent theft. Serialized prevention of forged pharmaceutical products includes security characteristics that customers can confirm, distinguishing between authentic and fraudulent items with the help of chain code-based transactions. People expose society to threats by forging trust in companies and providing poor-quality drugs. Numerous opportunities, including an anti-forgery medicine system and IPFS network, are used to protect against forgery in the Ethereum blockchain:

- Set ownership constraints for non-retail and retail drugs to prevent drug cloning.
- Create practical ACMS administration with Ethereum contracts using IPFS networks and the host system.
- Activate the initiative for small business firms.
- Analyze the outcome.

Protection from scams can be effectively achieved through ACMS, where blockchain code is generated to initiate transactions by the client. The next verification step involves a digital signature, which sends the order service and validation to assess the quality of product distribution. Pharmaceutical distribution aims to facilitate ledger systems, chain codes, and serialization, assigning serial numbers to pharmaceutical products for easier identification and differentiation. Preventing illegal access to the cloud can be achieved by exposing the system to threats and addressing delivery challenges. This impacts business operations and poses issues with drug tracking and traceability, corporate operations, patient safety, and regulatory compliance.

With sophisticated secured tracking and monitoring technology, commodities in transportation are delivered on time to the desired locations, allowing for undisturbed pharmaceutical business with medical system management [9]. To enhance the international delivery of medicine securely, an international database was established. Smaller companies should also benefit from the technology, whereas it offers great prospects for large pharmaceutical companies with protective measures for the protection of pharmaceuticals since they are valuable commodities. The pharmaceutical sector's safety and security are based on blockchain technology's cryptographic capabilities. Additionally, unauthorized drug modifications are prevented by stopping fraudulent pharmaceutical investors from changing medications and degrading their quality.

2.3.1 OPEN RESEARCH CHALLENGE IN THE PHARMACEUTICAL INDUSTRY

Recent technology in the pharmaceutical industry aims to verify the serial number throughout the entire supply chain system, improving drug traceability and quality control with the help of digital signatures as explained in Figure 2.3. However, various parameters such as blockchain codes, patient data, and data miners must be sent from the producer to the pharmacist to maintain constant quality.

2.3.2 PHARMA AND GENERIC DRUGS DATA DISPARITY

Data is crucial in the pharmaceutical industry as it enables experimentation for understanding and developing novel medications to treat various diseases. The pharmaceutical industry often faces data discrepancies as a major challenge when trying to cure specific diseases, and it holds the main responsibility for differentiating the data and storage [10]. Meanwhile, drugs with generic names are becoming more affordable, and the FDA has decided to expedite and simplify the introduction of generic drugs to the market.

2.3.3 RISING CUSTOMER EXPECTATION

Customers are becoming more sophisticated and aware as time goes on, and their expectations are rising. The pharma industry is responsible for effectively implementing commercial and therapeutic strategies to deliver services that meet customers' expectations in both underdeveloped and developed countries.

2.3.4 SUPPLY CHAIN AND DATA ANALYTICS

Maintaining the supply chain and data in the pharmaceutical industry is a major concern and a risky process. The pharmaceutical industry faces financial losses due to corruption and fake medications whenever the supply chain is compromised. According to a global analysis of the healthcare infrastructure, pharmaceutical businesses are predicted to lose 200 billion dollars due to the sale of fake medications.

FIGURE 2.3 Open research challenge in the pharmaceutical industry.

2.4 BLOCKCHAIN-BASED MODEL FOR SMART HOME NETWORK SECURITY

The rapid advancement and ubiquitous utilization of sensor, blockchain, and Internet of Things (IoT) technologies have provided greater potential for developing smart home systems [11]. However, due to rapid market expansion, some security challenges are incorporated with smart homes, and it gives a primary priority for people's attention and how directly it connects to everyday lives. These vulnerabilities could endanger information assets and seriously affect people's health and well-being, so the researchers are focused on reducing security threats such as fragmentation, firmware security, and unification of system architecture. Figure 2.4 illustrates the various research challenges and issues in smart home network security.

2.4.1 CONFIDENTIALITY AND INTEGRITY

The method employed in smart cities is a new technology that provides more sophistication and responsiveness and is already incorporated with blockchain technology. Instead of providing sufficient defenses, already it consists of several sensor nodes to provide security solutions with the help of defense mechanisms. A security breach is one of the main challenges in network security. The users in the network have become completely concealed as compared to fully traceable in blockchain technology. The transactions are often public and accessible to the user groups since cryptocurrencies have higher transparency. The surveillance of user behavior and the disclosure of individual's real identities could result from this, and the records could be used to facilitate data sharing. Therefore, true confidence should be improved.

2.4.2 STORAGE SECURITY

Maintaining and analyzing such data has become increasingly difficult in recent years due to the exponential amount of data produced by modern technology. Cloud service providers offer enormous amounts of storage and computational capacity. The risk of moving operations to cloud storage could result in significant delays with possible threats and attacks. Furthermore, the integrity of the data is impossible and dangerous in the global world where fraud or attackers are present among cloud service providers [12]. It is also necessary for a data owner to validate the authenticity of outsourced data. To this extent, it introduced a variety of distributed cloud storage mechanisms. Furthermore, these approaches are vulnerable to malicious attacks and single-point failure. To address the above issues, highly centralized blockchain storage solutions must be implemented.

2.4.3 ENERGY CONSERVATION

Energy conservation is an essential factor in smart cities and many consensus algorithms, like PoW and other metrics, give efficient computational power to sensor nodes for mining from one block to the next block [13]. However, the PoW consumes more power and time while calculating. However, the methodology is not acknowledged as being energy-efficient.

Even though the results are acknowledged with favorable findings, further research and investigation are necessary to improve the consensus algorithms because PBFT demands robustness and PoW security has not yet been thoroughly examined.

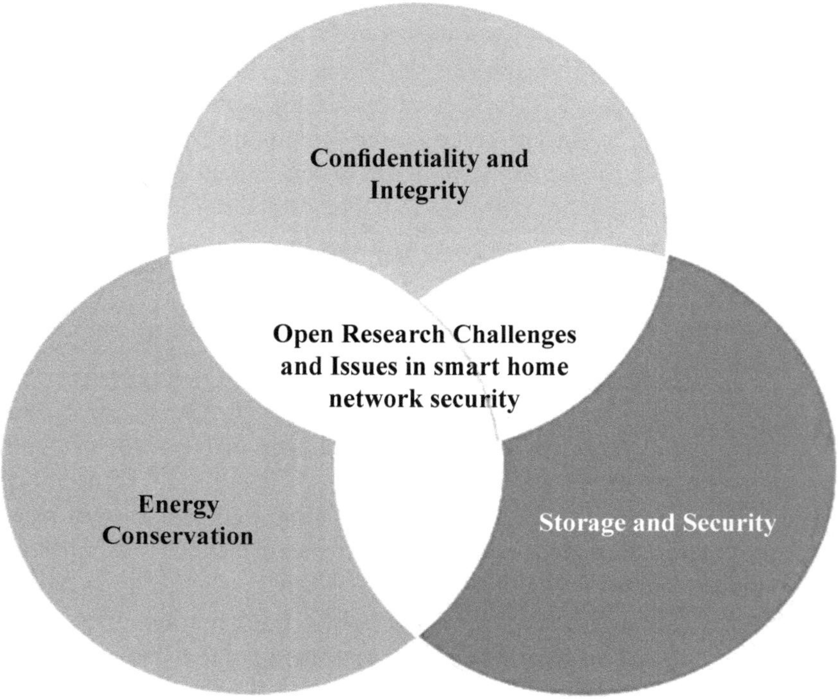

FIGURE 2.4 Open research challenges and issues in smart home network security.

2.5 BLOCKCHAIN WITH DEEP LEARNING TECHNOLOGY

Deep learning has gained tremendous growth in recent days, and it has the potential ability to make better decisions. However, operational integrity, data provenance, dependability, privacy, and accountability show a characteristic of insufficient tolerance in deep learning systems on central servers. And at the same time, single-point failure might arise when deep learning models are trained on data integration. Blockchain integration is categorized as a thematic taxonomy depending on a few factors, such as consensus protocol used on blockchain technology, deep learning models, data types, and deployment area. By installing these parameters, some of the problems hinder the integration of technology scalability, data exchange securities, the latency of transaction execution, support for platform integration, the volume of data collected from wearables and sensors, sustainable market mechanisms, and computationally expensive consensus procedures are going to highlight the opportunities and research challenges in deep learning models.

2.5.1 OPEN RESEARCH CHALLENGES IN BLOCKCHAIN WITH DEEP LEARNING

There are various challenges that exist in blockchain with deep learning models. Some of them are as follows:

1. **Scalability:** Deep learning applications significantly impacted the availability of various blockchain technology configurations and modifications in recent days. Efficient blockchain networks can increase the amount and speed of transactions multiple users generate. The objective of implementing deep learning-based services is to manage traffic in global blockchain networks, healthcare industries, cellular networks, congested highways, and other areas. Several issues will arise from the global adoption of blockchain, most of which are related to user demand for bandwidth, data rate, and the number of transactions generated by users [14]. By computing the processing and storage needs, the blockchain ledger should constantly grow by adding new blocks and transactions. To meet the anticipated needs of users, the number of block frames and transactions incorporated in the blockchain must be substantially reduced. To address scalability difficulties in present blockchain platforms, compression algorithms have been widely used. However, the current compression algorithms have a lightweight design with a high compression ratio, but they cannot fully deliver the optimal ratio for large-scale deployment.

2. **Data Security, Sharing, and Validation:** Private blockchain platforms utilize advanced encryption and decryption methods that authorize participants to exchange confidential data in the public healthcare sector, including physicians, patients, and nurses. Data security, stability, and transparency are more important

considerations in the IoT and wearable technology and expand exponentially in the healthcare and other network industries [15]. Secure data sharing between the participants can be facilitated by a multi-layer blockchain architecture that enables data fusion with enhanced analytical authorization for groups of users. Through a blockchain platform, users have been approved to verify passports and exchange security keys, which meets the requirements and industry standards for building a reliable and fault-tolerant system.

In vehicular technology, the blockchain and AI are used to collect vehicle data. However, it may be flawed or deployed improperly due to a lack of security. As a result, the information produced by these gadgets is frequently inaccurate, fraudulent, and unreliable. To use the metadata binding technique devices must calculate its accuracy and precision level together with the blockchain data. The deployed smart contracts can analyze this metadata to determine whether the data is accurate before saving it on the blockchain [16, 17]. Additionally, in such instances, a reputation-aware system may be useful for determining the degree of trust level for every participating individual based on their actions. It is possible to use smart contracts to ensure that only data from reputable companies is stored on the blockchain.

3. **Data Storage Capacity:** Deep learning can solve a wide range of real-world challenges, and one of the big issues is securely storing data. With the help of blockchain technology and deep learning, it is highly possible to store the data with high security, immutability, traceability, transparency, and accessibility, as mentioned in Figure 2.5 [18, 19]. The deep learning-based methodology is economically preferable because it systematically evaluates the blockchain platform's performance. An analysis of the blockchain's current structure provides potential structural improvements for solving real-world object recognition and classification issues across various fields; deep learning techniques also offer a more effective way to compress data [20]. As the blockchain increases in size, it significantly impacts how the network performs. Deep learning, along with blockchain methodologies, can be used to eliminate redundant data and compress the data. It is crucial to focus on designing a new algorithm to develop data compression.

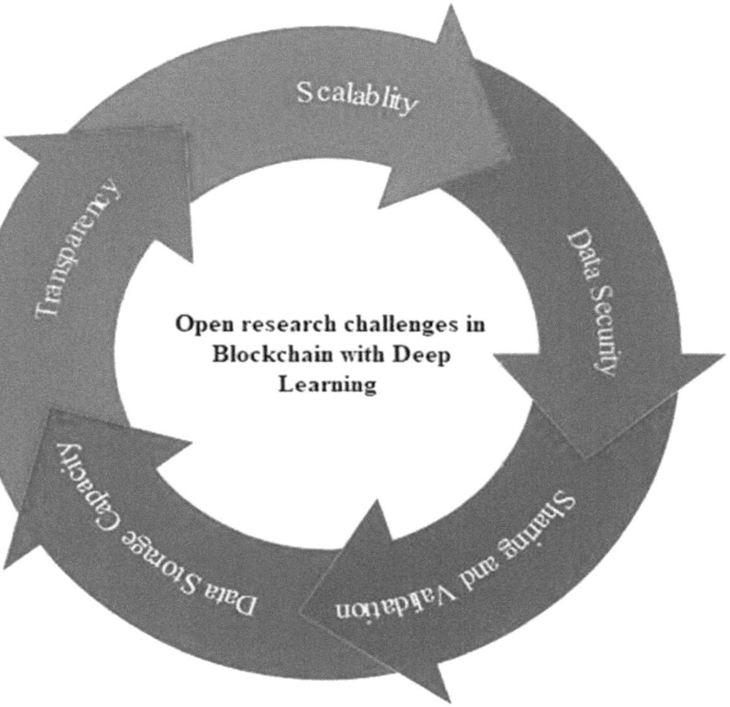

FIGURE 2.5 Open research challenges in blockchain with deep learning.

2.6 BLOCKCHAIN ORACLES

Blockchain interoperability is a novel aspect of the technology that is quickly gaining acceptance in many industries. However, blockchain networks behave like isolated islands that cannot link and share resources or information due to a lack of communication between the blockchains, and the execution of smart contracts is limited to on-chain only [21], with a focus on blockchain Oracles and how the blockchain ecosystem operates as intermediaries to integrate data sources.

2.6.1 OPEN RESEARCH CHALLENGES AND FUTURE DIRECTIONS OF THE ORACLE

1. **Trust and Security Issues:** Security and dependability are the primary advantages of blockchain, as it provides cryptocurrency-based

methods and consensus algorithms. On the other hand, Oracles, who are outsiders that provide data to the blockchain ecosystem, lack authenticity. Dealing with unauthorized data sources raises security concerns and difficulties. Typically, they do not offer strong assurances on the authenticity of the data, potentially introducing inaccurate or outdated data to the blockchain [22, 23]. Therefore, ensuring the accuracy and integrity of the data becomes impossible. Several design considerations have been implemented to guarantee the dependability and integrity of Oracles. The significance of blockchain lies in maintaining security and dependability to ensure the success of cryptocurrency approaches, but it is crucial to address issues by improving security and privacy-related design and operations of Oracles.

2. **Performance Evaluation of QoS:** Towards overall investigation, Oracle has not yet been thoroughly examined or analyzed to measure QoS performance and its fundamental characteristics. Even though several studies have been addressed and some performance parameters are to be evaluated in the blockchain-Oracles ecosystems such as integrity of data, throughput, transactional delay, flexibility, scalability, and all basic metrics that must be considered for future advancement.

2.7 BLOCKCHAIN TECHNOLOGY FOR SUPPLY CHAIN MANAGEMENT

Handling the movement of goods and services between various stakeholders is the responsibility of Supply Chain Management (SCM). The traditional SCM is modeled for a centralized approach, with all other sub-offices receiving instructions from the main office, which is handled at the central headquarters [24, 25]. Current SCM systems' primary challenges include security, transactional transparency, traceability, stakeholder engagement, product duplication, increased latency, corruption, and unreliability. On the other hand, blockchain and SCM technology manage data effectively and transparently while establishing trust. Moreover, researchers need to focus on transacting the payments directly to a party or with supply chain transaction permission with valid verification.

2.7.1 OPPORTUNITIES OF BLOCKCHAIN FOR SECURED SCM

Maintaining the integrity of products and procedures in a multi-stakeholder supply chain system constitutes a significant challenge. Data fragmentation, a lack of verifiable authenticity, and various protocol requirements apply to various distributions and processes that hinder the efficiency of many contemporary approaches [25, 26]. Blockchain has been characterized as an advanced technology among alternative solutions since it provides secured transparency and protection, data integrity, and the growth of stakeholder confidence [27]. Figure 2.6 depicts the Opportunities of Blockchain for secured SCM.

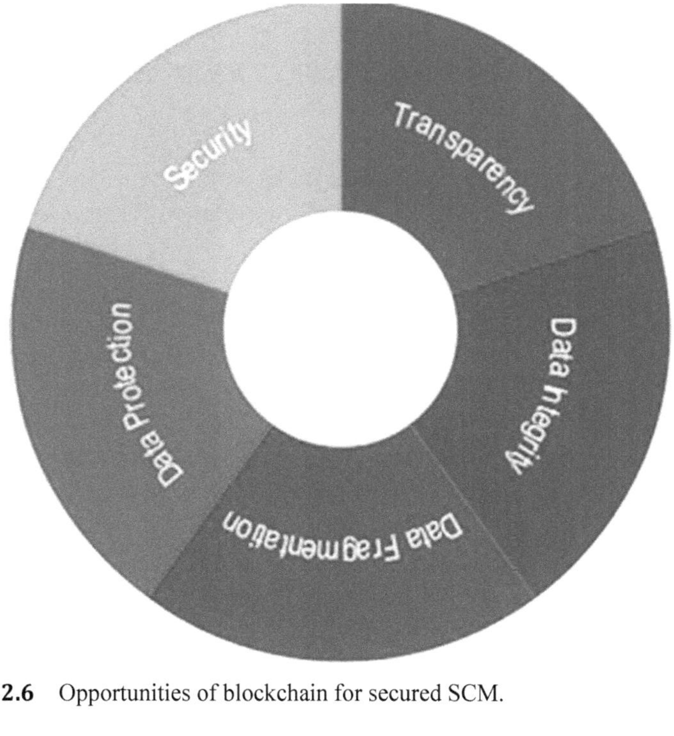

FIGURE 2.6 Opportunities of blockchain for secured SCM.

2.8 DETECTING FRAUDULENT IN HEALTH INSURANCE FUZZED BLOCKCHAIN AND AL

Health insurance (HIC) is now becoming more important in today's society due to the rise in medical complications. For individuals who cannot afford expensive operating costs, medical crises may be problematic. Health insurance assists individuals in acquiring medical care in the case of emergency services and offers financial security with the help of Blockchain technology. Health insurance may be threatened by safety, confidentiality, and scams, creating losses for individuals, businesses, and governments [28]. The utilization of fraudulent healthcare data has remained a central issue for both national authorities, and commercial businesses, and the development of technologies to identify fraudulent cases and payments has been increasingly important in this area over the past few years. Therefore, it is more important to resolve issues in blockchain and AI-based healthcare insurance.

2.8.1 OPEN RESEARCH CHALLENGE IN HIC BY DETECTING THE FRAUDULENT

There are several research challenges in AI-empowered HIC Fraud Detection as shown in Figure 2.7. A detailed overview of these challenges is in subsections.

2.8.1.1 CLASS IMBALANCE

HIC fraud classifiers are constructed using machine learning models that perform poorly due to class imbalance, which is a significant challenge in recent research. With more records than the minority class, the majority class uses a computer modeling approach to categorize the imbalances. It is more difficult to determine the minority class because it has fewer HIC records. As a result, a comprehensive and elaborate approach to the class imbalance issue must be put forward.

2.8.1.2 STANDARDIZATION OF DATA EXCHANGE RULE

No standard law or regulation pertains to the cross-border exchange of HIC data. Different governments have different privacy laws governing the trade of HIC data. The benefits of blockchain and AI-enabled HIC fraud detection systems while data sharing may be constrained when multiple authorities are involved in HIC data. Consequently, more research is needed to standardize the retrieval and storage of HIC data, as well as legislation and regulation.

2.8.1.3 ACCESSIBILITY OF THE DATASET

Public and private healthcare sectors should have increased accessibility to actual HIC data. Real-world HIC data sets have proven hard to come by because of issues with the law, privacy, and competition [29]. Future studies using open-source HIC real-world datasets are crucial for enabling other researchers to verify their findings. To obtain the suggested solutions employing actual healthcare insurance data, researchers need to work with the healthcare insurance industry.

2.8.1.4 HANDLING A GROWING NUMBER OF USERS

Monitoring the expanding user base in the HIC fraud detection framework powered by AI and blockchain is one of the challenging tasks. The need for the healthcare insurance industry is rising as a result of population increase for the benefit of patients. Integrating AI and blockchain-enabled frameworks in real-time scenarios gets harder as the framework's insurance subscribers increase. HI companies have the potential to combine with technologies like sensors and wearables that are employed in healthcare for increasing operational functionality, which is mentioned in Figure 2.7. HI providers can acquire real-time data on insurance subscribers, evaluate their premium costs, and identify HIC fraud for wearable technology applications. Future research should therefore focus on the data analytics and accessibility for blockchain HIC fraud detection systems.

2.8.1.5 LACK OF PROFESSIONAL TALENT

One of the most quickly developing technologies in the modern era is the combination of blockchain and AI. It takes highly qualified individuals who are familiar with intricate AI, cryptography, and advanced mathematics techniques to develop blockchain-based fraud detection apps. At the moment, only a small number of researchers and experts are focused on developing robust and adaptable blockchain and AI-based systems.

Due to a lack of qualified personnel, implementing fraud detection in AI and blockchain will be complicated.

2.8.1.6 IMPROVING SMART CONTRACT DEVELOPMENT

Executable Solidity programs known as "smart contracts" are loaded on Ethereum Virtual Machines and used to regulate behavior under agreements. Because smart contracts cannot be changed after they are distributed in the blockchain network, it is necessary to fix the software fault before it is done. To apply the smart contract effectively and reliably, research solutions should be reinforced.

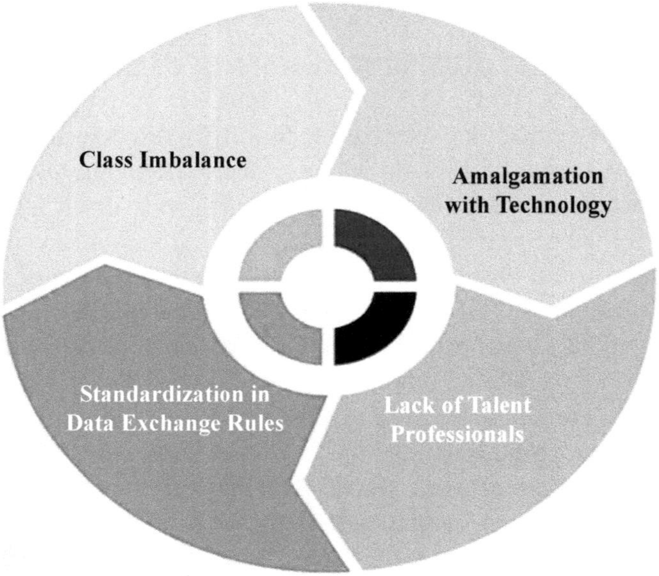

FIGURE 2.7 Open research challenges in AI-empowered healthcare insurance fraud detection.

2.9 CONCLUSION

Due to the involvement of several interconnected parties and ongoing complex processes and challenges in digital twins, cost management, risk assessment, and efficiency improvement need to be the focus. To analyze, anticipate, and optimize all processes before real-world implementation, one of the potential options to maximize efficiency is to develop a digital twin that mirrors every aspect of the underlying system or process. Furthermore, the research community must look into the research and deployment challenges. Digital twins are created by integrating AI and blockchain technologies to provide predictive maintenance and technological intelligence.

KEYWORDS

- **artificial intelligence**
- **blockchain technology**
- **digital twins**
- **healthcare**
- **risk assessment**
- **security**
- **technological intelligence**

REFERENCES

1. Sbaha, S., Rasheed, H., Raja, D., & Alma, O. (2023). Blockchain-based digital twins: Research trends, issues, and future challenges. *ACM Computing Surveys, 54*(11), 1–54.
2. Umamaheswaran, S. K., Prasad, O., Abdul, Z., & Vashistha, P. (2022). Major challenges and future approaches in the employment of blockchain and machine learning techniques in health and medicine. *Security and Communication Networks*.
3. Dhiviya, S., Malathy, S., & Kumar, D. R. (2018). Internet of things (IoT) elements, trends, and applications. *Journal of Computational and Theoretical Nanoscience, 15*, 1639–1643.
4. Padmaja, M., Shitharth, S., Prasuna, K., & Chaturvedi, A. (2022). Grow of artificial intelligence to challenge security in IoT applications. *Wireless Personal Communications, 127*, 1829–1845.
5. Yvanoff-Frenchin, C., Ramos, V., Belabed, T., & Valderrama, C. (2020). Edge computing robot interface for automatic elderly mental health care based on voice. *Electronics, 9*, 419.

6. Hussien, H., Yasin, S., Udzir, N. I., & Ninggal, M. I. H. (2021). Blockchain technology in the healthcare industry: Trends and opportunities. *Journal of Industrial Information Integration, 22*, 100–217.
7. Rajendran, S., Mathivanan, S. K., Jaygopal, P., Janaki, K. P., Malar, B., A., & Somanathan, M. S. (2021). Emphasizing privacy and security of edge intelligence with machine learning for healthcare. *International Journal of Intelligent Computing and Cybernetics*.
8. Bhattacharya, P., Tanwar, Bodkhe., & Tyagi. (2019). Blockchain-based deep-learning-as-a-service in healthcare 4.0 applications. *IEEE Transactions on Network Science and Engineering, 8*, 1242–1255.
9. Rajesh, D., Kadry, S., & Kang, B. (2022). Probit Cryptographic Blockchain for Secure Data Transmission in Intelligent Transportation Systems. *Journal of Internet Technology, 23*, 1303–1313.
10. Ammi, M., Alarabi, S., & Benkhelifa, E. (2021). Customized blockchain-based architecture for a secure smart home for lightweight IoT. *Information Processing & Management, 58*, 102–482.
11. Aruchamy, P., Sabeena, G., & Sowndarya, K. (2023). An artificial intelligence approach for energy-aware intrusion detection and secure routing in the internet of things-enabled wireless sensor networks. *Concurrency and Computation Practice and Experience*, 1–28.
12. Shafay, M., Ahmad, Salah., Yaqoob, J., & Omar. (2022). Blockchain for deep learning: Review and open challenges. *Cluster Computing*, 1–25.
13. Prieto González, L., Gomez, B., & Popa, de Amescua Seco, A. (2021). A survey on energy efficiency in smart homes and smart grids. *Energies, 14*(21), 72–73.
14. Weng, J., Weng, Zhang, Li., Zhang, & Luo. (2019). Deep chain: Auditable and privacy-preserving deep learning with blockchain-based incentive. *IEEE Transactions on Dependable and Secure Computing, 18*(5), 2438–2455.
15. Vazhuthi, P., Aruchamy, P., & Manikandan, S. P. (2023). A hybrid ANFIS reptile optimization algorithm for energy-efficient inter-cluster routing in Internet of Things-enabled Wireless Sensor Networks. *Peer-to-Peer Networking and Applications, 16*, 1049–1068.
16. Lu, Q., Xu, Liu., Weber, Z., & Zhang. (2019). BaaS: A unified blockchain as a service platform. *Future Generation Computer Systems, 101*, 564–575.
17. Bhaskar, K. B., Prasanth, A., & Saranya, P. (2022). An energy-efficient blockchain approach for secure communication in IoT-enabled electric vehicles. *International Journal of Communication Systems, 35*, 1–20.
18. Huang, H., Zhu, Xiao, P., Xiao, F., & Sun, X. (2020). A blockchain-based scheme for privacy-preserving and secure sharing of medical data. *Computers & Security, 99*, 102010.
19. Tan, L., Yu, K., Shi, N., Yang, C., Wei, W., & Lu, H. (2021). Towards secure and privacy-preserving data sharing for COVID-19 medical records: A blockchain-empowered approach. *IEEE Transactions on Network Science and Engineering, 9*(1), 271–328.
20. Gupta, R., Reebadiya, D., Tanwar, S., Kumar, N., & Guizani, M. (2021). When blockchain meets edge intelligence: Trusted and security solutions for consumers. *IEEE Network, 35*(5), 272–278.

21. Ezzat, S. K., Saleh, Y. N., & Abdel-Hamid, A. A. (2022). Blockchain oracles: State-of-the-art and research directions. *IEEE Access, 10*, 67551–67572.
22. Lo, S. K., Xu, Staples., & Yao. (2020). Reliability analysis for blockchain oracles. *Computers & Electrical Engineering, 83*, 106582.
23. Mohsin, A. H., Zaidan, M., Albahri., & Alsalem. (2021). PSO–Blockchain-based image steganography: Towards a new method to secure updating and sharing COVID-19 data in decentralized hospitals intelligence architecture. *Multimedia Tools and Applications, 80*(9), 14137–14161.
24. Agarwal, U., Rishiwal, T., Chaudhary, S., & Bokoro. (2022). Blockchain Technology for Secure Supply Chain Management: A Comprehensive Review. *IEEE Access*.
25. Wamba, S. F., & Queiroz, M. (2020). Blockchain in the operations and supply chain management: Benefits, challenges, and future research opportunities. *International Journal of Information Management, 52*, 102064.
26. Colicchia, C., Creazza, Noe., & Strozzi. (2018). Information sharing in supply chains: A review of risks and opportunities using the systematic literature network analysis (SLNA). *Supply Chain Management: An International Journal, 24*(3).
27. Kapadiya, K., Patel, G., Alshehri, Tanwar., & Sharma. (2022). Blockchain and AI-empowered healthcare insurance fraud detection: An analysis, architecture, and future prospects. *IEEE Access, 10*, 79606–79627.
28. Vacca, A., Di Sorbo, A., Visaggio, C. A., & Canfora, G. (2021). A systematic literature review of blockchain and smart contract development: Techniques, tools, and open challenges. *Journal of Systems and Software, 174*, 110–117.
29. Wang, C., Cheng, H., & Xiao. (2021). A survey: Applications of blockchain in the internet of vehicles. *EURASIP Journal on Wireless Communications and Networking, 2021*, 1–16.

CHAPTER 3

Blockchain Cryptic Chills: Paving the Way for a Futuristic Internet

GAGANJOT KAUR,[1] SUGANDHI MIDHA,[2] and MARIYA OUAISSA[3]

[1]*Project Manager, Iotasol Technologies, Ludhiana, Punjab, India*

[2]*Assistant General Manager, College Dekho, Chandigarh, India*

[3]*Professor, Moulay Ismail University, Meknes, Morocco*

ABSTRACT

Blockchain innovation is more established than Bitcoin. It is an essential central part of most cryptocurrency organizations. A decentralized, distributed, publicly accessible electronic ledger must maintain a highly permanent record of all recently confirmed transactions. Internet of Things (IoT) and blockchain advances are, by and large, heavily taken advantage of and utilized in many spaces, particularly for e-medical care. The blockchain is deliberately designed to be highly resistant to change. Blockchain addresses various issues by providing decentralized algorithms and storage for IoT data. Blockchain technology certainly offers modern prospects for the functionalities of smart contracts. The fundamental idea of a smart contract is the automatic execution of specific tasks upon the fulfillment of certain conditions. The supply chain is the biggest adopter of blockchain after crypto. Blockchain is also used in the insurance industry, fintech, energy, and medical records. Like blockchain, crypto is maturing, and because blockchain offers a permanent record, it is expected that soon, everyone will see much greater adoption as the ecosystem of third-party support around blockchain uses continues to expand. This chapter deals

Blockchain-Based Digital Twins: Research Trends and Challenges.
A. Prasanth, Vandana Sharma, Rajesh Kumar Dhanaraj, Pethuru Raj (Eds.)

with the foundation of Blockchain in different application areas in business focusing on handling data and transaction management. In order to provide the technical knowledge and implementation aspects of Blockchain, this chapter provides the detailed implementation of Ethereum.

3.1 INTRODUCTION – THE TRUTH ABOUT BLOCKCHAIN

Without the use of any middlemen or centralized authorities, peer-to-peer exchange of tangible and digital assets is made possible by blockchain technology. It was first created to promote using Bitcoin as a cryptocurrency for financial transactions. Although blockchain was initially developed as the technology behind Bitcoin, its potential benefits go far beyond digital money. Decentralized technology has made significant progress with the development of blockchain. Virtually all monetary transactions in the internet marketplace are processed by independent financial institutions, raising transaction fees, capping the accurate minimum purchase price, and opening more doors to fraud than trust [1]. Because it is an encrypted electronic payment system that does not rely on faith, blockchain was developed as a distributed technology to circumvent the fundamental shortcomings of this trust-based approach. Through hashing, timestamping, tree architectures, and consensus procedures, blockchain enables peer-to-peer transactions over decentralized networks, removing the need for intermediaries [2]. Moreover, it provides complete transparency about transaction history, makes available any subsequent transactions that have been placed since the initial one, and enables any organization to verify such transactions at any moment [3, 4].

Blockchain technology can help centralized companies address low security, high costs, and inefficient operations [5]. Blockchain technology enables peer-to-peer exchange of tangible and digital assets without intermediaries or centralized authorities. Initially developed to facilitate the use of Bitcoin as a cryptocurrency for financial transactions, the Bitcoin protocol served as the foundation for the first blockchain implementation. The double spending of digital currency was one of the issues that the Bitcoin Blockchain was created to address. It is very challenging to stop users from copying digital things in the current era. Despite having little success, the film and music businesses have been battling unauthorized downloading for years. If users can make fake copies whenever they want, digital money is not viable. The feat achieved by "Satoshi Nakamoto"—the alias of the creator(s) of Bitcoin—is enabling peer-to-peer trades of digital currency safely. The Bitcoin system establishes trust between strangers in a completely decentralized manner through cryptographic methods and financial incentives.

The ledger was encrypted and stored in a decentralized manner. There are two groups of transactions, one including those that still need to be completed and another containing those that still need to be included in the blocks of the main chain. To add to the distributed ledger, each block "aggregates" a set of recently processed transactions and assigns them a timestamp. The process of "mining," which involves intense competition and a lot of computer work, validates new transactions and produces new blocks. Like cryptocurrencies, various participants in an IoT system (nodes, gateways, and users) may not always trust one another while carrying out transactions [6]. Since it is a distributed ledger system, blockchain can accommodate the safety standards necessary for IoT programs. Public-key cryptography is used by blockchain technology to record transactions between several decentralized parties. Blockchain-based peer-to-peer networks may be used to mitigate the effects of a catastrophic failure at a central node.

3.1.1 CORE FUNCTIONALITY OF THE BLOCKCHAIN

To put blockchain technology to use, a P2P network consisting of all the nodes eager to do so must be set up. Two different keys, one public and one private, are assigned to each node in the network. A node's public key is used to encrypt data given to it by other users, while its private key is used to decode the data. For this reason, we use not just one but two separate keys: one for encryption and one for decryption. A user's public key serves as an address, whereas blockchain transactions are signed using their private key.

Communications encrypted with a particular public key may be decrypted only by the owner of the accompanying private key. Learn why asymmetric cryptography is so named here. After a node signs a transaction, it performs it and broadcasts it to its immediate peers. The transaction's authenticity and integrity are guaranteed by its unique signature. The signed transaction is received by the peers of the node that broadcasts it, and as soon as they have confirmed its validity, they retransmit it to other peers to disseminate it throughout the network. Particular nodes called "master nodes" organize and arrange the transactions distributed in this fashion and accepted by the web into a timestamped block. A Bitcoin transaction works in the following manner:

1. All network nodes share an identical copy of the Bitcoin ledger before the transaction. This displays the Bitcoin addresses and the corresponding balances, like all ledgers.
2. Each participant has a set of keys that relate to mathematics. Both public and private keys are used. The sender starts a transaction by sending some Bitcoins to the beneficiary's address.
3. After the sender signs the transaction using his private and recipient keys, it is broadcast to the whole network. No other crucial but the recipient's private one may be used to decode the payment; however, anybody can verify the transaction's provenance using the sender's public key.
4. Transactions are verified by a particular group of participants known as "Miners," who then compile them with all the other transactions into a "block."
5. Bitcoins that were sent now belong to the beneficiary, and accordingly, balances on the global ledger are updated.

During Bitcoin transactions, there are three types of blocks: those in the main chain, which include encrypted confirmations of previous transactions. Those related to the main chain but not included, for some reason, and orphan blocks. The first two sets form a directed rooted tree with back-to-the-root links, whereas the orphan sets stand alone. From a technical standpoint, the blockchain is a chain of hash pointers that may be used to create a linked list of past transactions, and the rooted tree eventually converges on the blockchain. Each block is marked with a unique cryptographic signature. Each participating network in a Bitcoin transaction manages its own private/public key infrastructure. (i) transaction irreversibility, the principle that a purchase or sale cannot be undone; and (ii) When cloning is out of the question, it is also impossible to create new currency by printing.

Blockchains can be divided into permissioned and permissionless blockchains. Any peer can connect to and disconnect from a permissionless blockchain as a reader or writer. Users can read the information, which is decentralized, so permissioned blockchains authorize limited readers and authors. They are controlled by a central organization, which determines who can read and write [7]. The business has promised blockchain. It is optional for the firm to use a solely public blockchain. Blockchain can be used for commercial purposes in a private, permissioned setting where participants are familiar with one another, and cryptocurrency is not required to establish transactions. A certain amount of authority is also allocated among the participants. The network must keep up the shared ledger.

The members of the permissioned network have access to the shared ledger, but only those transactions they are permitted to read are visible to them. A certain amount of restriction is placed on the blockchain network by the permissioned blockchain. The most popular method for managing keys for cryptography in blockchains is distributed public key infrastructure (PKI) based on blockchains [8]. PKI methods based on blockchains lack a trusted third-party or central access point. Furthermore, these methods provide transparency in the public system without presupposing any particular degree of trustworthiness on the side of the system's nodes or participants.

3.1.2 REQUIREMENT OF THE BLOCKCHAIN

Transparency can be achieved through sharing data and using blockchain technology. These methods clarify the public system without presupposing any particular degree of trustworthiness on the side of the system's nodes or participants. Some of the factors that make blockchain technology popular across numerous industries include the ones listed below:

1. Transparency is ensured. This is the only one who can change your blockchain data since the technology is open source. The immutability of blockchain-stored data increases the system's security, making it easier for early adopters to create and provide solutions.
2. Transaction costs are drastically decreased. On a blockchain, transactions between individuals and companies may be finalized without a trusted third party. The procedure moves more quickly because no middlemen are engaged in the deal.

3. Compared to conventional banks, which are limited by their business hours and established procedures, blockchain technology enables immediate settlement of transactions. Location in different parts of the globe also influences the lag time. Blockchain, on the other hand, doesn't have such restrictions. Therefore, it can settle transactions far more quickly.
4. Since there is no central data hub, it encourages decentralization. This makes it possible to authenticate certain transactions. Even if data is updated on numerous servers and confronts threats, a small quantity of data will be lost.
5. It monitors the flow of products, resulting in transparency. Additionally, this streamlines various other management procedures.
6. When abnormalities are found, it is always possible to return to the source, simplifying investigations and allowing for the quick execution of necessary steps. This results in quality control.
7. Human mistake is eliminated by blockchain technology since it stores data and prevents tampering. Accuracy is ensured when records are checked off as they go from node to node. Because of this, a sense of responsibility is developed.
8. Blockchain technology makes validating, signing, and enforcing complex and intelligent contracts simple.
9. By eliminating electoral fraud, blockchain technology improves voting transparency.
10. Blockchain technology's dependability is being considered for stock exchanges.
11. The supply of energy can be precisely monitored.
12. Blockchain technology promotes international peer-to-peer trade. Transactions with cryptocurrencies are quick, safe, and affordable. Astonishing products that sweep out the existing and competing products.
13. Blockchain helps slow market development leading to a dip and the technology uptake to establish the market.
14. By using blockchain technology, data becomes objective. In addition to maintaining data integrity, it can notify users when data has been changed. Even if a company's data is breached, it cannot

be used in any way that would undermine security. Therefore, the two do not conflict.

15. Using blockchain technology, we can verify the authenticity of devices in 15 different ways. Potentially eliminating the need for human contact, passwords might soon become obsolete. This is so that centralized architecture is not encouraged.
16. Non-repudiation is emphasized by digitally timestamping and signing each transaction. Even in the most recent version of the system, all historical data will be preserved in the log. It also results in traceability.

3.1.3 BLOCKCHAIN ARCHITECTURE

Like a traditional public ledger, blockchain consists of blocks that include transaction information. As seen in Figure 3.1, a blockchain is an example of a distributed ledger system. Each block has just one parent, and that parent's hash is included in the block header. Remember that the hashes of uncle blocks (the offspring of the ancestor block) The transaction details would also be added to the Ethereum ledger. There is no parent block to the first block on a blockchain, called the "genesis block." In subsections, we go deep into the workings of blockchain.

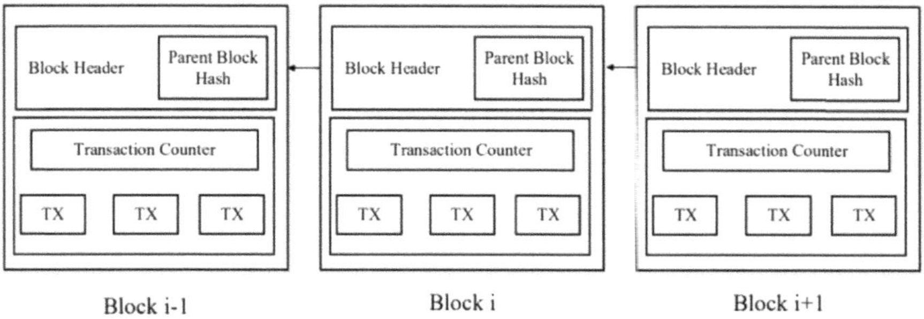

FIGURE 3.1 The blocks in this sample blockchain are linked together in a linear fashion.

3.1.3.1 BLOCK

It is clear from looking at Figure 3.2 that each block has both a header and a data section. The block header includes:

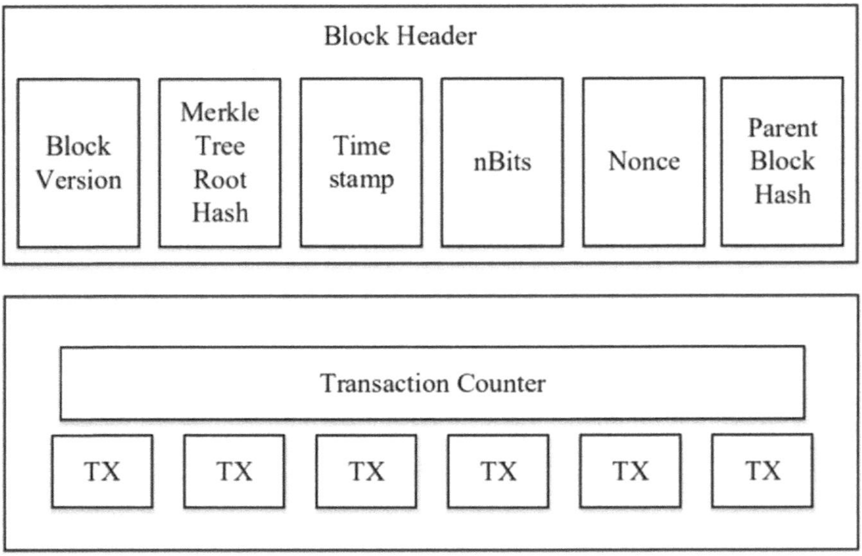

FIGURE 3.2 Block structure.

i. **Block Version:** Identifies the collection of rules for validating blocks.
ii. **Merkle Tree Root Hash:** Calculated by adding up each transaction's soup in the block.
iii. **Timestamp:** Time in seconds since January 1, 1970, at midnight UTC.
iv. **nBits:** Maximum acceptable block hash goal.
v. **Nonce:** 4 bytes that increase value by 1 with each hash.
vi. **Parent Block Hash:** 256-bit hash value that references the previous block.

A transaction counter and the data encoding those transactions make up the bulk of the block. How many transactions may fit within a block is determined by the block size and the size of each transaction. To verify the legitimacy of transactions, blockchain relies on asymmetric cryptography. Asymmetric cryptography-based digital signatures are utilized in suspicious settings. Next, we provide a quick example of a digital signature.

3.1.3.2 DIGITAL SIGNATURE

Private and public keys are associated with each user. They were sharing the signed transactions with everyone on the network. Digital signatures generally consist of two steps: signing and verification. Let's say user Alice is interested in contacting user Bob. The signature method involves Alice encrypting some information with her private key and sending the encrypted and unencrypted copies to Bob. Bob could quickly determine whether the data has been altered. Digital signature algorithms are crucial to blockchains, and the most widely used method of digital signature generation is based on elliptic curves.

3.1.3.3 KEY CHARACTERISTICS OF BLOCKCHAIN

In a nutshell, the following are some of blockchain's most distinguishing features:

1. **Decentralization:** Traditional centralized transaction solutions add additional expenses and have the core servers sluggish since the central trusted agency must validate each transaction. Unlike the old centralized system, blockchain technology removes the requirement for a centralized server or trusted third party to process transactions. Blockchain uses consensus techniques to ensure data integrity over a decentralized network.
2. **Persistency:** A transaction's validity can be checked quickly, and trustworthy miners will not approve incorrect transactions. Whenever a new block is created on the blockchain, it can never be undone or removed. Invalid transaction blocks might be quickly uncovered.
3. **Anonymity:** A user's identity must not be revealed to communicate with the blockchain; instead, they will use a randomly created address. It is important to remember that owing to an inherent restriction, blockchain cannot provide absolute privacy protection (details will be discussed in Section 3.4).
4. **Auditability:** Using the Unspent Transaction Output architecture, the Bitcoin blockchain keeps information on user balances: An unutilized transaction must be referenced in any given transaction.

When a new blockchain transaction is added, any unspent transactions it references automatically transition from unspent to spent. Therefore, confirming and keeping tabs on your transactions would be simple.

3.1.3.4 TAXONOMY OF BLOCKCHAIN SYSTEMS

Today's blockchain infrastructures may be broken down into three distinct types. A public blockchain lets everyone see and verify past transactions and work together to establish a consensus. On the contrary, a consortium blockchain would have just chosen a few nodes to participate in the consensus process. Private blockchains restrict participation in the consensus process to nodes belonging to a single company.

Since only one entity has authority over the nodes in a private blockchain, the network is highly concentrated. Because a subset of nodes would be used to reach an agreement, the consortium blockchain proposed by numerous organizations is not decentralized:

1. **Consensus Determination:** Any participant node in a public blockchain may forge a consensus. Only some of the nodes in a consortium blockchain verify the addition of each new block. On the other hand, private chains are managed entirely by one entity, which also has the power to set the rules for reaching a consensus.
2. **Read Permission:** The public can see all transactions made on a public blockchain; however, this is only sometimes the case on a private or consortium blockchain.
3. **Immutability:** Due to the distributed nature of the ledger, altering a transaction on a public blockchain is very difficult. In contrast, tampering with transactions is less likely in private or consortium blockchains due to the smaller number of users.
4. **Efficiency:** Because the public blockchain infrastructure is so vast, the dissemination of blocks and transactions takes some time. This results in low transaction throughput and significant delays. Consortium blockchains and private blockchains could function better with fewer validators.

5. **Centralized:** Each blockchain type has different characteristics: While the public blockchain is entirely decentralized, the consortium blockchain has a degree of centralization, and the private blockchain, which a single firm governs, has no decentralization.
6. **Consensus Process:** Public members can participate in the blockchain's consensus process anywhere globally. Permissioned blockchains, such as consortium and private blockchains, are distinct from public blockchains in several ways.

Since everyone worldwide may join a public blockchain, numerous users and vibrant communities often exist. Every day, new public blockchains appear. Consortium blockchain has a wide range of potential uses in the business world. For the time being, Hyperledger is developing blockchain frameworks for business networks. To aid in the creation of consortium blockchains, Ethereum has provided the necessary technology.

3.1.4 RELATED WORK

Here are two brief reviews of projects that showcase work that does similar things using IoT and blockchain technology.

An approach to Recharging Electric Vehicle Batteries using the Ethereum Blockchain to facilitate exchanging batteries in a system for recharging electric car power packs [9]. It developed a network of connected devices based on the Ethereum blockchain. It was suggested that a "rich-thin-client" approach be used instead. Wealthy clients are machines that have the power to store the whole blockchain ledger and take part in mining operations. Thin clients are low-powered computers that can only keep the blockchain data directly applicable to them. This allows low-powered IoT nodes to participate in the blockchain network without shouldering the burden of the network's cumbersome consensus procedures.

Connecting Battery-Powered IoT devices to a distributed ledger may interact with a blockchain network without becoming nodes themselves. A gateway node is a computer on the blockchain network that receives messages from computers that do not have blockchain software installed. Messages are processed at the gateway before being sent to the blockchain. This proof-of-concept gateway shows that even small IoT devices may interact with the blockchain network without becoming nodes in the network itself. According to Eyal [10], blockchain is used in the banking and healthcare industries to store and secure user account information and patient records. Users are no longer restricted to their physical locations for data access because of the decentralized nature of the network.

Blockchain technology may help businesses save money on transaction fees, improve efficiency in their core operations, and increase the transparency and ease with which they are compensated for using their intellectual property. A competitive edge for businesses is possible [11]. Bitcoin and other virtual currencies backed by blockchain technology provide a potential solution to the funding issues faced by enterprises. To provide one example, cryptocurrency may help businesses who wish to switch to accepting just electronic payments. Automation in electronic transaction management accounting enhances internal and external financial company execution control [12].

Because information is essential to every business, blockchain allows for secure data storage and seamless sharing [13]. Blockchain's advantages as a distributed and encrypted ledger make it a good fit for managing digital assets at various businesses [14]. Due to the immutability of the blockchain, decentralized data storage eliminates the need to trust a single entity with your information. Instead, it makes it available to users all over the globe. A firm might utilize the blockchain to store data, making it more transparent, secure, and less vulnerable to manipulation.

The distributed ledger technology known as blockchain might completely alter the logistics industry [15]. Improvements in supply-chain provenance have been made possible by recent implementations of blockchain technology and the IoT [16]. Suppose information about a product's journey from the factory to the consumer is stored on the blockchain. In that case, businesses will have a much easier time tracking the origins of their goods and keeping an eye on the quality of the finished goods.

Blockchain is often used to establish company collaborations and inter-organizational business procedures, which businesses may use to create smart contracts. Companies may use smart contracts on blockchains to conduct automated transactions without requiring human verification. Under smart contracts, firms, for instance, may automatically submit tax returns [17].

The authors in Ref. [18] discuss the securities industry's possible use of blockchain technology. They delve into pressing concerns about the regulation and security of European financial markets. Their paper is less of a how-to guide and more of a place to air their thoughts on the possibility of such technology being used in the securities business within the context of European standards' social and legislative restrictions. As a result, they refrain from drawing any firm conclusions and conclude that putting such technology into practice would be challenging at best and so is unlikely shortly.

The term "smart city" refers to city planning that emphasizes three key elements necessary for a city to flourish via the collaborative production and consumption of products and services. The three main variables are humans, technological advancements and implementations, and urban coordination [9]. Most people live in urban areas, and another 2.5 billion people are projected to move by 2050. Consequently, numerous problems, such as pollution, traffic, and trash, will likely intensify. At the very least, issues like fraud and liability caused by incompetent service providers may be eliminated using blockchain technology. Third, a blockchain platform supported by the local government and geared toward documenting and organizing all of these interactions might provide the assurance needed for such a smart city to materialize if the potential rewards outweigh the potential risks.

3.2 BLOCKCHAIN ARCHITECTURE

This technology of Blockchain is the backbone of the present and future information-sharing business world. This is the technology behind cryptocurrency, secured healthcare, Supply Chain Management, and so on. Before understanding Blockchain's architecture, let us deeply understand its components and parameters.

3.2.1 BLOCKCHAIN COMPONENTS AND PARAMETERS

If we break up the word Blockchain, we conveniently get two words: Block and Chain. Block is a data structure responsible for holding transactions and related data. Why are we using a Block data structure? Why don't we store data in one prominent data structure as in the cloud? In the cloud, we store data on one single centralized server. In Blockchain, we have multiple small blocks to store the data. The primary concern here is due to security reasons. The security aspect of Blockchain has been explained in detail in the later part of this and the next section.

Transactions generated at a particular point in time are gathered in one block. After some time, new transactions generated would be concluded in new blocks. Publishing transactions in new blocks is a continuous process and keeps on happening forever and ever. Each time a transaction is generated, it comes and sits inside a block and is published in the main blockchain ledger, as depicted in Figure 3.3.

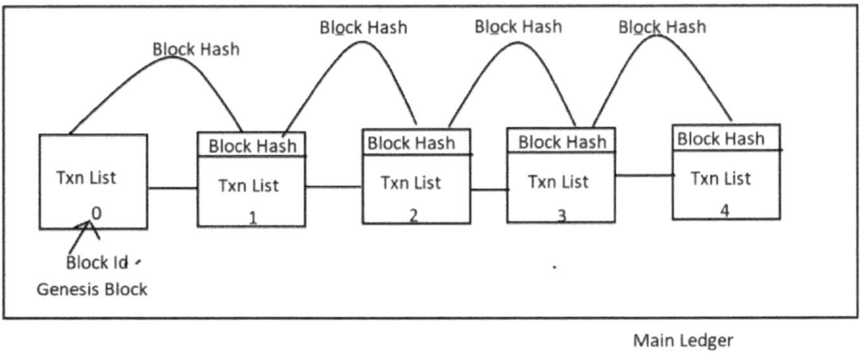

FIGURE 3.3 Chain of blocks.

A block holds the transactions, and when we hear the word "transaction," most transactions that come to mind are financial transactions like funds transfers, financial asset transfers, etc. However, this is different from what happens in the blockchain. In this, along with financial transactions, even information is part of it. Anything or everything that happens in the blockchain environment is a transaction. Everything is initiated as a transaction, stored as a transaction, and recognized as a transaction.

Blocks are recognized in the blockchain network uniquely with a block hash. It has one more critical parameter, which is the block ID or number. The very first block of the blockchain network is the Genesis Block. The block number is the position of the current block from the Genesis Block. The block hash is a unique hash provided to each block. Specific hashing algorithms are used for calculating the hash for each block. Each block has a unique set of data which generates a unique hash. Take an example of the Ethereum blockchain network [12–14], as depicted in Figure 3.4. It is one of the famous block explorers of blockchain.

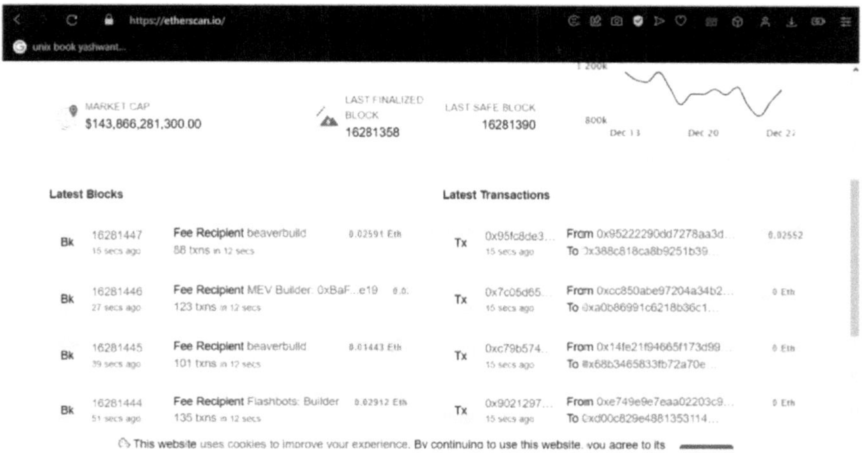

FIGURE 3.4 Ethereum blockchain explorer.

The left side of the explorer shows the latest blocks created, and the right side shows the latest transactions. Transaction parameters are listed in Figure 3.5. Let us observe the parameters of one of the blocks in the explorer, as depicted in Figure 3.6. Block Height defines the Block Number, and Block Hash uniquely represents the Block. Block Hash is never stored in the same block where it is developed. It is stored in the next Block Header as "Previous Block Hash (PBH)." It creates a linkage between all the blocks. This is precisely how they are chained together. These blocks are linked with the parameter "PBH," and this process is continuous and forever. That is why it is termed blockchain.

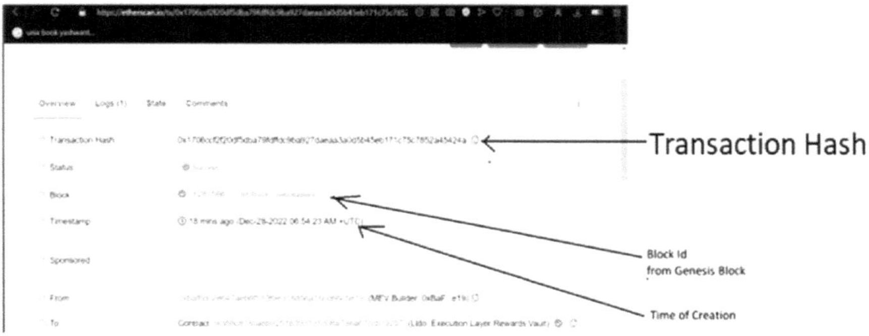

FIGURE 3.5 Transaction parameters.

FIGURE 3.6 Block parameters.

Ledger is not unique, but it is a collective term given to a chain of blocks connected. One can compare the Ledger with the collective term, Humans. Humans are a group of people. Each individual in that group possesses their own identity and features. Similarly, a Ledger is a group of Blocks. Each block is unique in itself.

3.2.2 BLOCKCHAIN NETWORK TOPOLOGY ARCHITECTURE

To understand the Blockchain architecture [4], let's study it in complete form in a Centralized, Decentralized, and Distributed Network.

3.2.2.1 CENTRALIZED NETWORK ARCHITECTURE

In this architecture, everyone is dependent on a single centralized node. This single node handles data availability, data accessibility, and data upgradability. For example, consider a real-time application such as a Banking organization; in this case of a network, there is no flexibility for users to make decisions. Everything is dependent on the single central node. It acts as a single central decision-making point [15] as depicted in Figure 3.7.

Let us take another real-time example of an e-commerce web application (CW) such as Flipkart, Amazon, etc., where multiple sellers list their products on the commerce web application CW1. 1000s and 1000s buyers of CW1 associated with the site to purchase products of CW1 sellers. All sellers and buyers of CW1 are dependent on CW1 owners. CW1 owners have complete control over the data displayed on the application. CW1 can only update the value of products, Quantity of Products, etc. CW1 holds the ledger copy. It becomes the critical node in the network and the most vulnerable too.

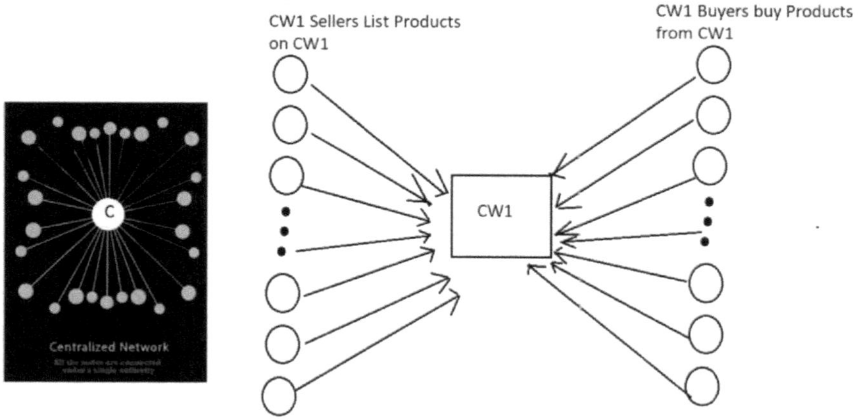

FIGURE 3.7 Centralized network architecture.

3.2.2.2 DECENTRALIZED NETWORK ARCHITECTURE

In the decentralized environment, CW1 acts as one of the critical nodes. There will be multiple nodes. Let us consider five sellers who are maintaining a copy of the ledger. All of them collaboratively work together in managing the data. As each one holds a copy of the ledger, even if one or two nodes go down, we have other nodes that can provide the data, as depicted in Figure 3.8. However, no one can provide the data if CW1 goes down in a centralized network. Consider a situation where if the Bank Server goes down, it notifies its customers that it is down. At that time, no customer could access the account. So mainly, if the server needs any maintenance time, it's done at night, so customers do not get impacted.

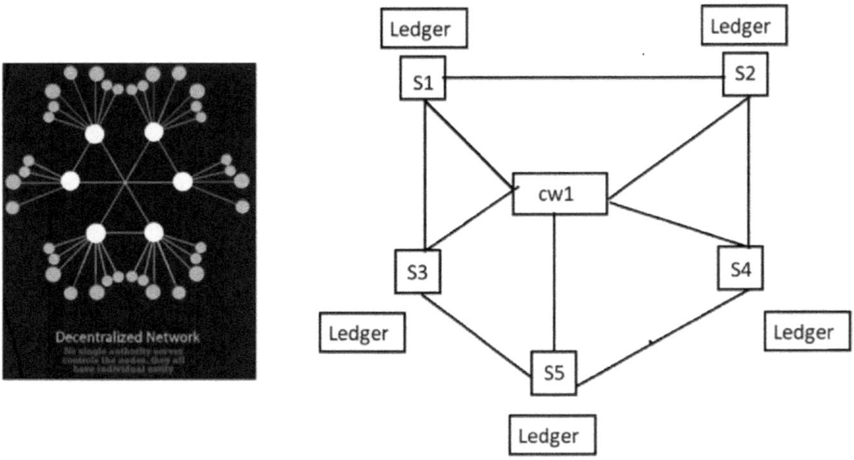

FIGURE 3.8 Decentralized network architecture.

A second important aspect in a decentralized environment is that CW1 is the network initiator but does not have complete control over the network. It acts as one of the entities responsible for ensuring data upgradability. All of them collaboratively work together. Sub-node customers are connected to each seller. One of the seller nodes and customers is depicted in Figure 3.9. Customers are the sub-nodes as they need help understanding the complexity of the network. CW1, in this case, loses power and ownership. It is the only source to exchange information.

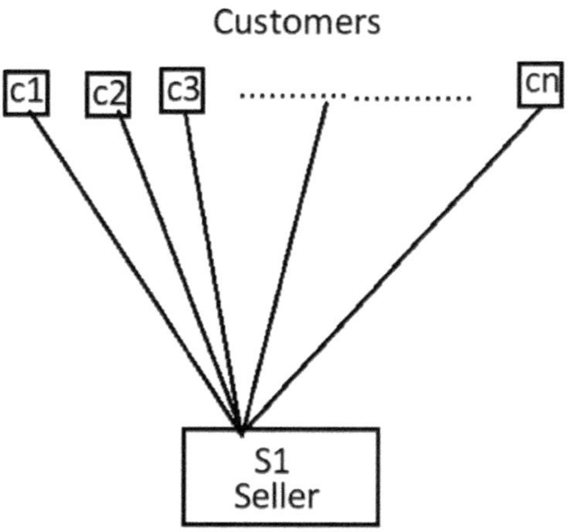

FIGURE 3.9. Seller and customer relations.

3.2.2.3 DISTRIBUTED NETWORK ARCHITECTURE

It is a closed compact network. Everyone is connected without depending on anyone absolutely. Anybody can view or transfer the data. Referring to Figure 3.10, white nodes hold a copy of the ledger. Peers are responsible for gathering, sending, and receiving the data. Nodes N1 to N4 are responsible for data availability and storing a copy of the data within them. The same copy of data is available at all four nodes, ensuring trustability, transparency, and security in the network. Multiple nodes storing the same data undoubtedly increase data redundancy, but this is important when constant availability is required.

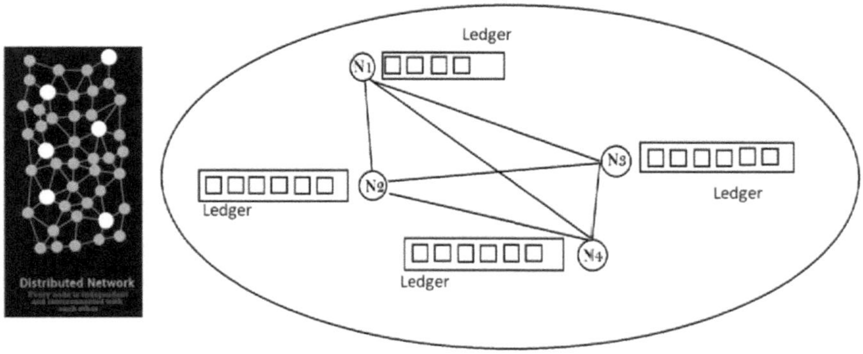

FIGURE 3.10 Distributed network architecture.

3.3 BLOCKCHAIN-BASED NETWORK APPLICATIONS AND THEIR SOLUTIONS

A question always comes to mind about what additional things blockchain brings to the world apart from existing ones. The answer to this is the utmost level of:

- Trust;
- Transparency;
- Security.

Blockchain is upskilling these three factors. There is no beyond level after this grade. The level of these three factors is unmatched by any other technology. Figure 3.11 is not new but shows the network topology followed by the blockchain environment.

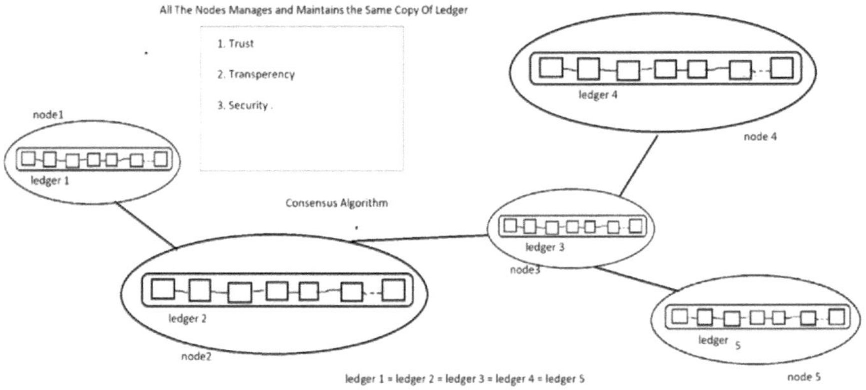

FIGURE 3.11 Network topology blockchain environment.

Trust and transparency are straightforward to identify and believe in the blockchain environment [3]. Each node maintains an exact copy of the ledger, which brings trust and transparency among all the nodes in the network. Let us understand the security [5, 6] aspect with the help of an application example of a chain of retailers and distributors network of clothing brands, as shown in Figure 3.12. The whole clothing brand network, which consists of manufacturers, distributors, and retailers, wants to ensure that everything is transparent. A clothing brand is the main initiator of the whole network.

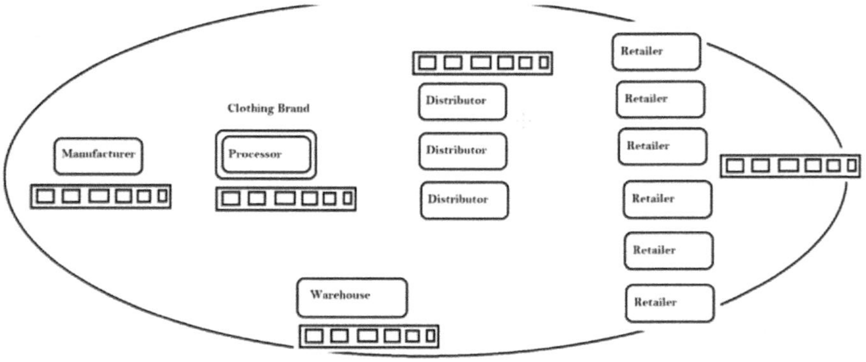

FIGURE 3.12 Clothing brand – an application of blockchain.

In the case of a centralized use case, the clothing brand would have been a vital player in ensuring data availability and maintenance. In a decentralized environment, all network participants are essential in supply chain management—each node in the network stores a replica of the ledger. The same piece of data as a clothing brand gets redundant at each other node in the network. Is there any better transparency than this? Not everybody can see what is happening in the network.

Now let us come to the security part of it. Even clothing brands want to tamper with the data, 51% approval is required to change the data. A single node cannot tamper with the data. In Figure 3.12, 12 nodes are there. At least the consent of seven nodes is required to tamper with the data. Once again, to prove the same fact, let us take a centralized environment where a single node works as a Single Point Shared/Private Server. It is easy for hackers to bypass the firewall, so they can easily hack and transfer data. Compare it with the blockchain environment. If a hacker wants to transfer the data to a particular block, consent, and changes at all the blocks at different locations must be changed. Nodes are unaware of what other nodes are available in the system, then how can a hacker determine the same?

First and foremost, a hacker needs to find out where exactly these different nodes are situated; if not all, but at least 51% of nodes, it should be able to find. Then only tampering would be possible. Multiple blocks allow the fetching of data quickly, preventing the system from being derailed by external hackers [9]. For some reason, let us assume the hacker can tamper with the data at node 1. However, is that enough to change the data at all places? Once again, 51% of the sites should have updated data to apply the change at all the locations.

Let us 00take another drastic situation where a hacker finds the location of 51% of nodes. However, one more rule ensures ultimate security in the blockchain environment. Change has to be done within the time frame of two continuous additions of blocks, i.e., within ΔT time, as depicted in Figure 3.13. Technically, it is an impossible and complex task to achieve. Let us see now how transaction flow happens in the Blockchain network. A detailed flow description is defined in Figure 3.14.

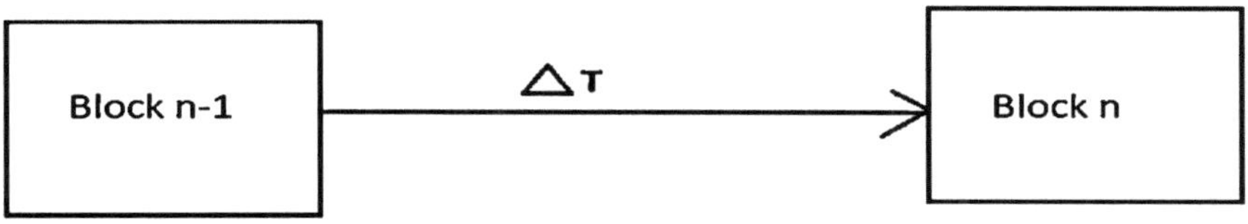

FIGURE 3.13 Time frame ΔT.

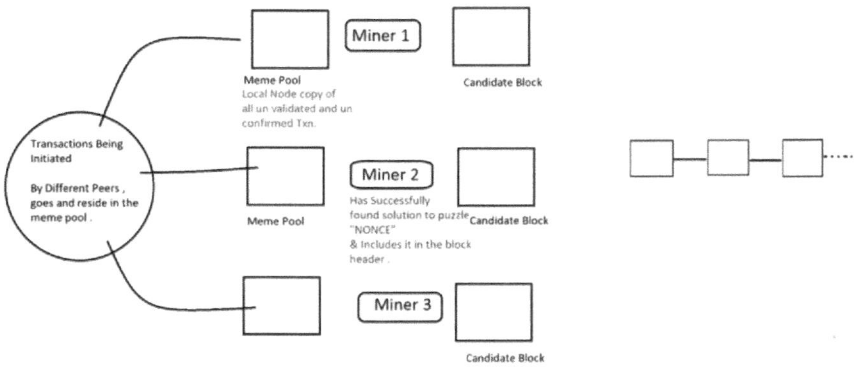

FIGURE 3.14 Transaction flow.

All the transactions in the blockchain are part of the Mempool. A Mempool can be compared to the local storage area of the miner responsible for holding unvalidated and unconfirmed transactions (Tans) from where the miner picks them. A core responsibility of the miner is to validate a transaction and create a block. Miner is a type of node responsible for creating blocks out of transactions. Miner validates them and, in turn, will receive some reward. Miner puts it in its block, which is known as a candidate block. It is not an actual fundamental block updated in the central ledger. All the candidate blocks will be competing with each other. And at the end, the block that wins the race will be updated in the final central ledger.

In reality, thousands and thousands of miners are there. In Figure 3.14, let us assume Miner 2 finds the solution to the Puzzle. Each and every time, the Consensus Algorithm launches the Puzzle. That particular Puzzle has to be solved by independent miners. Every miner will go into a competition to solve it. The miner who solves it first will be allowed to launch/publish its block in the central ledger. For example, in this case, Miner 2. The solution to the Puzzle is known as "NONCE." NONCE value is updated in the header of each block. Miners are not operated manually, but these are automated system applications that are independently capable of solving the Puzzle.

The consensus algorithm may be termed the heart of any blockchain environment, also known as a crucial component of the blockchain network. Each and every blockchain network needs a protocol or rule required to be defined, which can bind participants or network members together who are involved in the blockchain network. So, when a centralized authority is missing in any network, we find another alternative to a central authority. Consensus algorithm acts as that term in the blockchain. The consensus algorithm answers the following questions in the blockchain network.

- Who can create the transaction?
- How is the transaction validated?
- How are transactions approved?
- How are blocks generated?
- How would consensus be achieved?
- How can multiple network participants approve or agree to a specific fact/rule?

Specific rules and regulations are required if we want no one to derail the system. These rules and regulations will take care of everything. There is no dependency on anyone. Everyone has to follow a group of rules/scenarios. How consensus achieves this without a central authority. The network behaves smoothly with an extensive blockchain network. To understand the concepts, let us take an example where different applications are created or designed to target a particular operating system (OS) [16–18]. Without an OS, a system is hardware with no capability. The OS takes care of all the functions of the system. OS sub-layers take care of system hardware and its applications. Similarly, the Consensus Algorithm takes care of all the processes in the blockchain network.

The consensus Algorithm is efficient where thousands and thousands of miner nodes compete for a block to publish in a ledger. Every few seconds, the consensus is launching a new puzzle. If the number of miners is more, many transactions are generated. Multiple miners pick one transaction. Each transaction owner puts some reward for the transaction [19]. The more the reward, the more the interest of miners. And faster would be the transaction validation.

3.4 CHALLENGES AND THE ROAD AHEAD

While blockchain has enormous promise, it currently needs to be improved due to several issues that prevent its widespread adoption. We list some significant difficulties and promising developments.

3.4.1 SCALABILITY

When more and more transactions are added to the blockchain, managing it becomes more cumbersome. To verify blocks on the blockchain, every node must keep a copy of every transaction ever made, regardless of whether or not the funds from which it came are still available. It is slow to build because its blocks are enormous, and each one takes a long time to set up; the Bitcoin blockchain cannot perform real-time transactions. Currently, only around seven transactions can be processed per second on the network. Miners prioritize transactions with a high fee; many low-value transactions may be delayed because of the limited size of blocks.

To solve the blockchain's scalability issue, several solutions have been offered. These solutions may be broken down into two groups:

1. **Storage Optimization of Blockchain:** Due to the increased difficulty of running an actual copy of the ledger on a single node, Bruce introduced a groundbreaking method for managing Bitcoin transactions in which the network automatically deletes obsolete data. All non-empty addresses have their balances stored in an account tree database. A lightweight client could help solve this issue. Alternatively, a new scheme called Ver Sum was developed to enable lightweight clients to exist. Ver Sum facilitates the delegation of intensive calculations over significant inputs by lightweight clients. By comparing computing results across different servers, it guarantees accuracy.
2. **Redesigning Blockchain:** The bitcoin-next generation (NG) proposal was made where the idea is to divide the standard block into two parts: the fundamental block, which is in charge of choosing a new leader, and the macroblocks, which is responsible for recording individual transactions. Time is separated into several eras according to the procedure. At the beginning of each epoch, miners must conduct a hash to produce a new fundamental block. Distributing macroblocks to other nodes is the node's job that generates the critical block. Bitcoin-NG also developed the technique of the longest chain when macroblocks do not matter (heaviest chain). So, to solve this problem, a new blockchain is developed that does away with the need to prioritize block size above network security.

3.4.2 PRIVACY LEAKAGE

Using public and private keys in blockchain technology suggests that privacy may be somewhat preserved. Users can conduct secure financial transactions using their public key without revealing personal information. As illustrated, however, blockchain cannot provide transactional privacy due to the public nature of all transaction values and public key balances. In addition, recent research shows that it is possible to correlate Bitcoin transactions and expose the user's personal information. This holds even if the users are behind a NAT or firewall. IP addresses may be linked to their pseudonyms according to a technique introduced by Biryukov. Each client has a unique collection of connected nodes that serve as its fingerprint. However, the origin of a transaction may be tracked by learning this set. Several different approaches have been suggested to increase the privacy of blockchain, broadly falling into two categories:

Blending. Users' addresses on a blockchain are anonymous. However, since many users use the same address repeatedly, it is still possible to connect addresses to actual people. A mixing service is a service that allows users to remain anonymous by sending money to several recipients. Alice, who has address A, and lives at B, wants to send some cash to Bob. Alice and Bob's connection may be exposed if she conducts a direct transaction with input address A and output address B.

For this reason, Alice could give Carol a monetary gift who worked as a go-between. Carol's transfer to Bob has several inputs and many outputs. Bob's home address is one of those included in the output. This makes it more difficult to reveal Alice and Bob's relationship. However, a dishonest third party may purposefully reveal private information about Alice and Bob. Carol may not send Bob's money but instead to her own house. When using Mix currency, there is a simple method to avoid unethical behavior.

The user's requirements, such as the amount of money and the date of the transfer, are encrypted using the intermediary's private key. Moreover, if the go-between did not hand over the promised cash, anybody could easily confirm that the middleman had been deceived. Despite this, theft still needs to be stopped once it is discovered. The CoinJoin protocol uses a mixing server to randomly reorder the output addresses to avoid theft. *Coin Shuffle* is an address shuffling algorithm that uses decryption mix nets inspired by CoinJoin.

Anonymous. A zero-knowledge proof is utilized in Zerocoin. Instead of verifying a transaction's digital signature, miners must ensure that the coins in question are on an allowlist of acceptable currency. Payments are delinked from originating transactions to prevent payment origin from being analyzed as part of a transaction graph. Although the location and quantity of payments are hidden, they are revealed. To solve this issue, Zerocoin currency was suggested. In Zerocash, the zero-knowledge succinct non-interactive arguments of knowledge (ZK-SNARKs) are used. The total sum of transactions and the total value of users' cryptocurrency holdings remain anonymous.

3.4.3 SELFISH MINING

A blockchain may be attacked by a group of selfish miners working together. For example, the network remains susceptible even if a fraction of the hashing power is used to cheat. Selfish miners only share their blocks with others and hold on to them until certain conditions are met, at which point the private branch is made public. All miners would eventually accept the private branch as the new public chain since it is longer than the present one. In the interim, between now and when the private blockchain is made public, the honest miners spend their time and energy on an inactive fork, while the selfish miners are mining the private chain with no competition. Therefore, greedy miners often get more rewards.

Many more attacks have been developed to illustrate that blockchain is not so safe, and they all rely on selfish mining. To increase the profitability of obstinate mining, miners may combine mining operations with network-level eclipse assaults in a non-trivial way. One of the tenacious tactics' miners use is "trail-stubbornness," in which blocks are continually mined even if the private chain falls behind. However, there are situations when the advantages outweigh the costs by 13% compared to a non-trail-stubborn alternative. Selfish mining is not the only way to make money in the cryptocurrency industry; as shown, other, more lucrative approaches may be taken by smaller miners.

There are some benefits, although they are minimal. It also shows that the percentage of computational resources an attacker needs to profit from selfish mining is under 25%. Heilman suggested a new way for reliable miners to choose the best action, which might help alleviate the selfish mining issue. Sincere miners would choose more recently mined blocks if marked with random beacons and timestamps. However, timestamps may be easily forged. Time constraints mean that the network must generate and verify each block; Zero Block expands on this fundamental technique. Selfish miners cannot outperform the system's predicted reward in Zero Block.

3.4.4 BLOCKCHAIN: THE ROAD AHEAD

There will be difficulties in the development of any new technology. It is not like blockchain is a special case. There is a constant churn of efforts and activity in this space, all aiming to improve blockchain technology. This open and decentralized technology is excellent for trusting transactions, social connections, and business. Two of the numerous obstacles to the widespread use of the technology are its emphasis on transparency and openness among decentralized players. These issues may be resolved with the help of blockchain-deployed Depp since these applications allow for trustworthy financial dealings.

With everything you have learned so far, I hope you will closely examine the difficulties encountered by those working on every layer of the blockchain's infrastructure, from creating decentralized applications to improving the protocol itself. Here, we will go through some of the nice-to-haves for blockchain applications that must be more crucial to their functioning. Pay close attention to these characteristics as you plan and implement Depp's. This section thus overviews these features, difficulties, current remedies, possible prospects, and future directions.

- Decentralized identity management;
- Consensus among decentralized participants;
- Scalability, privacy, security, and confidentiality;• Public, private, and permissioned blockchain networks;
- The scientific research behind blockchain concepts.

3.5 CONCLUSION

Blockchain technology has established itself as a critical component of most cryptocurrency organizations and is increasingly being adopted by various industries for data and transaction management. The integration of Blockchain with the IoT has opened up new opportunities in areas such as e-healthcare, while the use of smart contracts has provided advanced functionalities. As Blockchain technology continues to mature, its adoption is expected to increase significantly, with industries such as production supply chain, fintech, energy, and healthcare leading the way. This chapter provided a comprehensive foundation for Blockchain technology in different business application areas, with a focus on data and transaction management, and a detailed implementation of Ethereum that serves as a valuable resource for understanding the technical aspects of blockchain.

KEYWORDS

- **blockchain**
- **cryptocurrency organizations**
- **cryptography**
- **healthcare**
- **internet of things**
- **selfish mining**
- **supply chain**

REFERENCES

1. Kottursamy, K., Banupriya, S., & AlZubi, A. (2023). A novel blockchain architecture with mutable blocks and immutable transactions for enhanced scalability. *Sustainable Energy Technologies and Assessments, 58*, 1–19.
2. Yli-Huumo, J., Ko, D., Choi, S., Park, S., & Smolander, K. (2016). Where is current research on blockchain technology? –A systematic review. *PLoS One, 11*(10), 1–27.
3. Fernandez-Carames, T. M., & Fraga-Lamas, P. (2018). A review on the use of blockchain for the internet of things. *IEEE Access, 6*, 32979–33001.
4. Hawlitschek, F., Notheisen, B., & Teubner, T. (2018). Be limits of trust-free systems: A literature review on blockchain technology and trust in the sharing economy. *Electronic Commerce Research and Applications, 29*, 50–63.
5. Rajesh, D., Kadry, S., & Kang, B. (2022). Probit Cryptographic Blockchain for Secure Data Transmission in Intelligent Transportation Systems. *Journal of Internet Technology, 23*, 1303–1313.

6. Shantha, R., Mahender, K., & Jenifer, A. (2022). Security analysis of hybrid one-time password generation algorithm for IoT data. *AIP Conference Proceedings, 2418*, 1–10.
7. Elsts, A., Mitskas, E., & Oikonomou, G. (2018). Distributed ledger technology and the internet of things: A feasibility study. *BlockSys'18: Proceedings of the 1st Workshop on Blockchain-enabled Networked Sensor Systems*, 7–12.
8. Herbert, J., & Litchfield, A. (2015). A novel method for decentralized peer-to-peer software license validation using cryptocurrency blockchain technology. *Proceedings of the 38th Australasian Computer Science Conference (ACSC 2015)*, 1–9.
9. Sun, J., Yan, J., & Zhang, K. Z. K. (2016). Blockchain-based sharing services: What blockchain technology can contribute to smart cities. *Financial Innovation, 2*(1), 1–9.
10. Eyal, I. (2017). Blockchain technology: Transforming libertarian cryptocurrency dreams to finance and banking realities. *Computer, 50*(9), 38–49.
11. Felin, T., & Lakhani, K. (2018). What problems will you solve with blockchain? *MIT Sloan Management Review, 60*(1), 32–43.
12. Zadorozhnyi, Z. M. V., Murayskyi, V. V., & Shevchuk, O. A. (2018). Management accounting of electronic transactions with the use of cryptocurrencies. *Finance Credit Activity Problems Theory Practice, 3*(26), 169–177.
13. Novikov, S. P., Mikheenko, O. V., Kulagina, N. A., & Kazakov, O. D. (2018). Digital registry of professional competences of the population drawing on distributed registries and smart contracts technologies. *Biznes Informatika Business Inform, 46*(4), 43–53.
14. Dutra, A., Tumasjan, A., & Welpe, I. M. (2018). Blockchain is changing how media and entertainment companies compete. *MIT Sloan Management Review, 60*(1), 39–43.
15. Treiblmaier, H. (2018). The impact of the blockchain on the supply chain: A theory-based research framework and a call for action. *Supply Chain Management: An International Journal, 23*(6), 545–559.
16. Bhaskar, K. B., Prasanth, A., & Saranya, P. (2022). An energy-efficient blockchain approach for secure communication in IoT-enabled electric vehicles. *International Journal of Communication Systems, 35*, 1–20.
17. Vishnevsky, V. P., & Chekina, V. D. (2018). Robot vs. tax inspector or how the fourth industrial revolution will change the tax system: A review of problems and solutions. *Journal of Tax Reform, 4*(1), 6–26.
18. Tranquillini, A. (2016). Blockchain yes, blockchain NO: An outsider (non-IT expert) view. *Journal of Securities Operations & Custody, 8*(4), 287–291.
19. Chaudhry, N., & Yousaf, M. M. (2018). Consensus algorithms in the blockchain: Comparative analysis, challenges, and opportunities. *2018 12th International Conference on Open-Source Systems and Technologies (ICOSST)*, Lahore, Pakistan, 54–63.

CHAPTER 4

Open-Source Platforms for Implementing Blockchain-Based Digital Twins

V. KARPAGAM,[1] P. V. KAVITHA,[1] and P. V. DEEPA[2]

[1]*Department of Artificial Intelligence and Data Science, Sri Ramakrishna Engineering College, Coimbatore, Tamil Nadu, India*

[2]*Department of Electronics and Communication Engineering, Arunachala College of Engineering for Women, Kanyakumari, Tamil Nadu, India*

ABSTRACT

A digital twin is a virtual software system that simulates the workings of an actual object or process. The majority of digital twin information is centralized and does not adequately support data management security, integrity maintenance, or trustworthiness and precision of time-consuming processes. It is imperative that blockchain be integrated with digital twins to address these restrictions. With blockchains, digital twin data can be shared across secured connection systems with confidence in terms of both accuracy and transmission speed. Blocks of data that track network transactions make up blockchains. These blockchains are maintained on distributed ledgers and are incorporated into the chain as new blocks. Several open-source frameworks of Blockchain such as Hyperledger Fabric, Ethereum, Corda, Quorum, Solidity, Geth, Remix, Mist, Solium, Truffle, Parity, DApp Board, Embark, MyEtherWallet, etc., were explored. Digital Twin open-source tools such as Ansys Digital Twin

Blockchain-Based Digital Twins: Research Trends and Challenges.
A. Prasanth, Vandana Sharma, Rajesh Kumar Dhanaraj, Pethuru Raj (Eds.)

Builder, Ditto, Kafka, AWS Digital Twin, AWS IoT TwinMaker, Twinbase, etc., were discussed. Also, various open-source platforms such as Dovetail, EtherTwin, SmartTwin, Remix, Solidity, TIBCO Cloud™ Live Apps, Watson IoT Platform, etc., for building Blockchain-based Digital Twin frameworks were discussed.

The chapter discusses the conceptual framework of blockchain-based digital twin and its significance in the industry sector. It explores the various open-source platforms for implementing blockchain-based digital twins. This chapter will be useful for researchers, academicians, and industry practitioners to understand the amalgamation of blockchain-based digital twins.

4.1 INTRODUCTION

A promising technique called the "digital twin" can significantly improve distributed manufacturing by delivering current functional data representations of actual assets, assisting in decision-making, and reducing potential dangers in such systems. Dispersed manufacturing systems must also work together to anticipate hazards and arrive at consensus-based decision-making [1]. However, due to centralized attack and connection, data interoperability, authentication, and scalability suffer from a single failure.

To mitigate the above-mentioned challenges, the adoption of Blockchain technology with Digital Twin enables safe data sharing and interoperability among numerous stakeholders, including manufacturers, suppliers, service providers, and users. Blockchains are digital databases or ledgers that consist of growing lists of records, named as a chain of blocks, each containing a list of transactions or data entries.

Blockchain and Digital Twins alter industry-oriented applications and guarantee data management security. Blockchain and Digital Twin technologies can deliver useful data and a host of advantages for decision-making and data management without the aid of a central authority. Furthermore, Blockchain-based technologies provide Digital Twins with security and integrity. As a result, merging Blockchain with Digital Twins improves corporate efficiency and legitimacy. Blockchain-based Digital Twin has the potential to change several sectors due to the growth in market share and requirements, and it also aids enterprises in broadening their horizons. Businesses can gather pertinent data about their products and processes by fusing blockchain technology and digital twins. Additionally, it motivates companies to produce more genuine and distinctive goods, giving them a competitive edge over rivals who might use analogous procedures for their goods. Indeed, combining digital twins with blockchain technology can greatly improve security, data privacy, and solution credibility in terms of product design and process outcomes. Although blockchain enables them to keep information, digital twins are created using conventional technology.

4.2 INTRODUCTION TO DIGITAL TWIN

Digital twin explores the simulated representation of an actual object or thing [2]. A thing can be a car, an engine, a building, or a flyover. The data can be gathered from sensors connected with these actual assets and converted into a virtual system. Because of digital twins, people can view important information about how an actual object performs in the real world. The addition of various cutting-edge domains like Intelligent Systems, Data Science, the Internet of Things, Sensors, Simulation, and Connectivity has significantly enhanced the sophistication and capabilities of digital twins [3].

4.2.1 FEATURES OF DIGITAL TWIN

1. **Connectivity:** It is the basis of a digital twin. It establishes a link between the actual element and its digital counterpart. Sensors provide the connectivity of physical objects by acquiring, integrating, and communicating data through different integration techniques.
2. **Homogenization:** Data homogeneity is both the result and the enabler of digital twins. It allows data to be separated from its actual form.
3. **Re-Programmability:** Digital twins provide programmability automatically via artificial intelligence techniques, sensors, and predictive analysis.
4. **Digital Traces:** These are left by digital twin technology. In the event of a machine failure, the trails can be used to locate the source of the problem.

5. **Modularity:** It refers to the design and customization of products and industrial modules. Modularity added to functional models allows producers to fine-tune machinery and models. Digital twin features are represented in Figure 4.1.

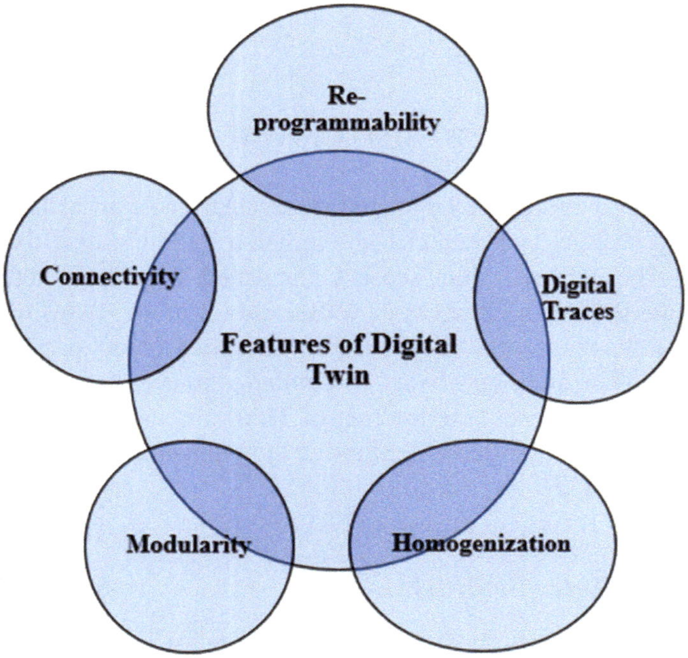

FIGURE 4.1 Digital twin features.

4.2.2 *WORKING OF DIGITAL TWIN*

A digital twin contains three main components. They are:

1. **Past Data:** It includes historical information, precise systems, and complete processes.
2. **Present Data:** Sensor data, outputs from production platforms, and pertinent information from systems along the distribution chain are all included.
3. **Future Data:** This is predicted data and insights from input and machine learning models.

The following are three working steps of digital twin [4]:

- **Step 1: Data Gathering and Simulation:** The data is athered and then simulated into a virtual model. Locally or on a distributed cloud, the gathered information is stored.
- **Step 2: Applying Cognitive Insights:** When simulation data is received, parameters are applied to tangible assets and merged for precise virtual representation. The data is reviewed to determine its variations and tolerances. To better understand the dynamics of the information being presented, natural language processing, machine-to-machine communication, acoustical video analytics, and other approaches are used.
- **Step 3: Dynamic Recalibration:** The environment around that particular gadget is dynamically recalibrated using the gathered information and cognition. This impacts the phases of design, build, and operation. The performance of tangible assets is enhanced by integrating data into both real-world and digital representations.

The process of digital twins is illustrated in the given Figure 4.2.

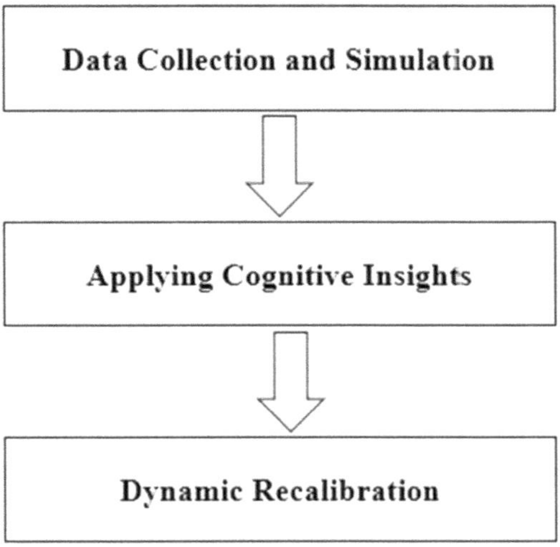

FIGURE 4.2 Digital twin workflow.

4.2.3 APPLICATIONS OF DIGITAL TWIN

The below are examples of how digital twin applications have made an impact across many industries [5]:

1. **Manufacturing:** Digital twins are applied in the manufacturing industry to speed up product development, design customization, shop floor performance enhancement, and preventive maintenance [6].
2. **Retail:** At well-known shopping malls and individual stores, digital twins are utilized to mimic and improve the consumer experience in the retail sector.
3. **Automotive:** Virtual representations of connected automobiles are frequently created using digital twins. To find potential issues before production or when the car is put on the road, automotive companies simulate and analyze the production phase.
4. **Healthcare:** By virtualizing healthcare services, digital twins assist healthcare providers in enhancing patient care, efficiency, and effectiveness. It attempts to increase individualized care and operational efficiency of healthcare processes [7].
5. **Smart Cities:** Digital twins can assist smart cities in achieving more social, economic, and environmental sustainability.

4.2.4 OPEN-SOURCE PLATFORMS – DIGITAL TWIN

1. **Ansys Digital Twin Builder:** It helps to implement full digital prototypes of real-world systems. These can be used to manage a product or asset's complete lifecycle. By scheduling maintenance around predictive approaches that improve accuracy with real-world testing and reaction, you can dramatically boost efficiencies over time. By having access to this data, engineers can maximize productivity while preventing unplanned downtime and reducing operating costs—all with IoT systems that stand on their connectivity.
2. **Ditto:** Eclipse Ditto's overarching objective is to create a foundation for digital twins. The coordination between physical

equipment and their virtual counterparts is the main focus. For instance, you may use weather information to improve a twin that knows its position. Consider an automobile as an example: it sends Ditto location data such as longitude and latitude every five minutes and travels from one spot to another. The weather information can be requested for the area where the automobile is now parked from a weather service in the interim. Ditto receives this data as feedback. As a result, the twin is always informed about the weather in the area where the automobile is currently parked.

3. **Kafka:** It is commonly used as a major event streaming platform to build scalable and dependable digital twins and digital threads for real-time streaming sensors. It provides an important event streaming platform for real-time communication, processing, and integration.

   Digital twin characteristics and their relationship to Kafka:

   i. **Connectivity:** Real-time scalability is provided by Kafka Connect for connectivity to IoT interfaces, cloud services, and big data. Other middleware and IoT systems are not in competition with the Kafka ecosystem; rather, they are complementary.

   ii. **Homogenization:** Or the actual decoupling between producers and consumers, is one of Kafka's key benefits. The administration and enforcement of schema via various technologies (JSON Schema, Protobuf, Avro, etc.) promote data awareness and standards.

   iii. **Reprogrammable and Smart:** Due to its programmability and intelligence, domain-driven design, and concern segregation (DDD) have become the de facto standard for microservice architectures. Versioning, A/B testing, and canarying should be used when launching new decoupled applications.

   iv. **Digital Traces:** These are used by the distributed commit log known as Kafka. Events are immutable, added, and preserved for as long as desired (possibly indefinitely with retention time = –1).

v. **Modularity:** Kafka's infrastructure is adaptable and modular. It includes client programs written in various languages such as Python, Scala, Go, Java.NET, C++, and others, as well as components like Kafka brokers, Schema Registry, Connect, REST Proxy, and others. With this level of adaptability, you can mix Kafka components with any other IoT solution to create the right Digital Twin architecture for your edge, hybrid, or global scenarios.

4. **AWS Digital Twin Projects:** AWS collaborates on digital twin initiatives with a lot of businesses. The following are a few of them:

   i. **Carrier:** It is a provider of cold chain and building solutions, leveraging AWS to develop its digital solutions quickly. The business uses AWS IoT services to build carrier.io, a shared services platform. In order to mimic its physical systems, it also leveraged AWS technology for asset modeling as well as the construction and integration of digital twins.

   ii. **INVISTA:** Specialty materials are utilized in a variety of industries, including clothes, automobiles, and computers. INVISTA is a Koch Industries company that specializes in these materials. To give employees a complete digital perspective of its assets and data, it collaborated with AWS to create digital twins of its manufacturing operations.

   iii. **John Holland:** One of Australia's top integrated infrastructure and real estate firms is John Holland. It was able to produce construction digital twins as part of a digital transition, giving managers a digital representation of their projects. AWS collects operational data for monitoring performance, the environment, claims, and historical information.

5. **AWS IoT Twin Maker:** It improves operations and performance by building digital twins of actual systems using AWS IoT TwinMaker. It duplicates physical structures including buildings, factories, manufacturing facilities, assembly lines, and industrial machinery with AWS IoT TwinMaker. For the purpose of creating 3D representations of your systems, import already 3D models into

AWS IoT TwinMaker, such as CAD and BIM files. It performs the following with AWS IoT TwinMaker:

- Improve the efficiency of the building;
- Increase production output quickly;
- Boost machinery performance;
- Look for and fix process irregularities;
- Keep an eye on and improve building conditions.

6. **Twin Base:** It is a Git-based Digital Twin Web server that is static and open source. It saves digital twin documents in a Git repository, and end users can retrieve the documents from the web server to be displayed using a web browser or client library [8].

4.3 INTRODUCTION TO BLOCKCHAIN

Blockchain technology is a distributed, immutable database that monitors asset and business transaction processes. A material asset, such as a home, money, vehicle, or plot, or an intangible asset, such as a patent, intellectual property, copyrights, and brand awareness, can be monitored by a blockchain network. This network can record and trade almost anything of value, reducing risk and lowering costs for parties during transactions.

In other words, Blockchain preserves data and information in a digital format without the need for a central authority. Data is gathered from numerous sources and organized into blocks on a Blockchain. Additionally, after they are full, these blocks link with subsequent blocks after including the sets of information. Thus, this network of data links and chains is referred to as a blockchain. Additionally, it stores the data and uses cryptography to link the blocks together.

4.3.1 FEATURES OF BLOCKCHAIN

Blockchain consists of the following features [9]:

1. **Decentralization:** The term "decentralization" describes a blockchain network that is not controlled by a central authority or intermediary. A blockchain, on the other hand, runs on a globally

dispersed network of nodes (computers) that each keeps a copy of the complete blockchain.

2. **Immutability:** This is a blockchain that refers to the characteristic of data being permanent and impervious to tampering once it is added. Data that has been stored in a block and uploaded to the blockchain can no longer be changed or removed. Consensus techniques and cryptographic hashing are used to achieve this.
3. **Consensus:** The network's users must reach a consensus on its validity using a consensus mechanism. Only authentic and valid transactions are added to the blockchain as a result of this agreement.

The features of Blockchain are depicted in Figure 4.3.

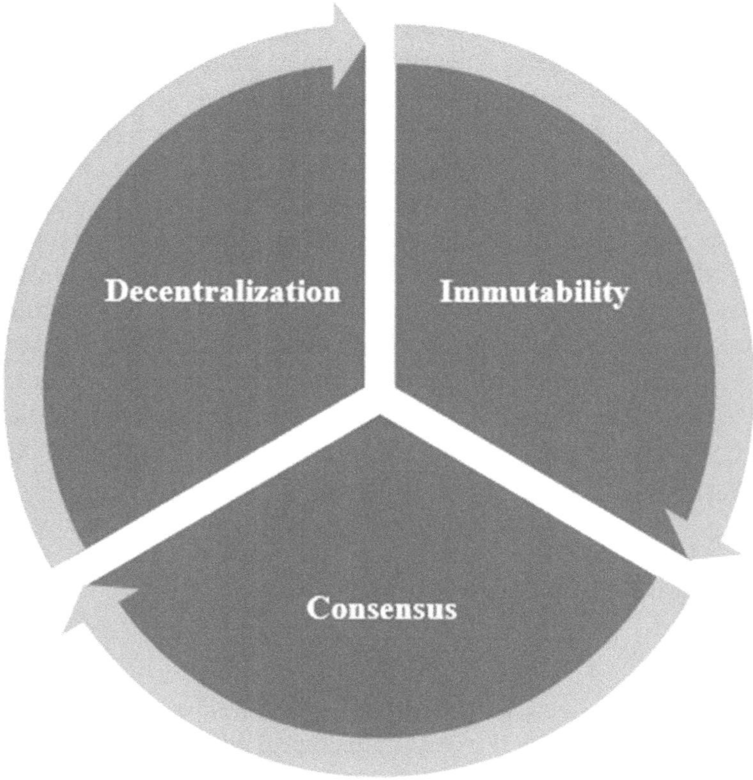

FIGURE 4.3 Blockchain features.

4.3.2 COMPONENTS OF BLOCKCHAIN

The key workings of Blockchain are as follows:

1. **Distributed Ledger Technology:** The system uses a shared and distributed ledger to record transactions one at a time in order to prevent duplicates.
2. **Immutable Records:** It prevents any data in the shared ledger from being transformed, changed, or tampered with.
3. **Intelligent Contracts:** It alludes to the set of guidelines for accelerating transaction times. Additionally, it runs and stores data in the blockchain. It establishes rules and guidelines for any transfers, including conditions.
4. **Public Key Encryption:** It serves as the fundamental building block of Blockchain. It makes use of public key encryption, a cryptographic method, to guarantee secure communication, confirm identities, and offer data integrity. Public Key encryption uses two cryptographic keys such as a public key and a private key [10].

4.3.3 WORKING OF BLOCKCHAIN

Blockchain works by the following steps. They are explained as follows:

1. **Step 1:** Record the Transaction: Every transaction is recorded as a block of data. It also covers the transfers and transactions of both tangible and intangible assets.
2. **Step 2:** Gain Consensus: The transaction was authentically recorded on the distributed blockchain network. The agreement rules differ based on the type of network, although they are often established at the outset of the network.
3. **Step 3:** Link the Blocks: After the parties have agreed, the transactions are then saved on the blockchain in blocks. The block contains the transactions and a cryptographic hash. The hash connects the blocks. If the block's data is modified intentionally or unintentionally, the hash value will change. The method of recognizing the data change is provided by this change in the hash

value. Therefore, the chains and blocks are always linked and immutable. Blockchain verifies all the blocks that are connected with each other.

4. **Step 4:** Share the Ledger: The system shares the latest copy of the ledger to each user [11].

The workflow of the blockchain process is illustrated in Figure 4.4.

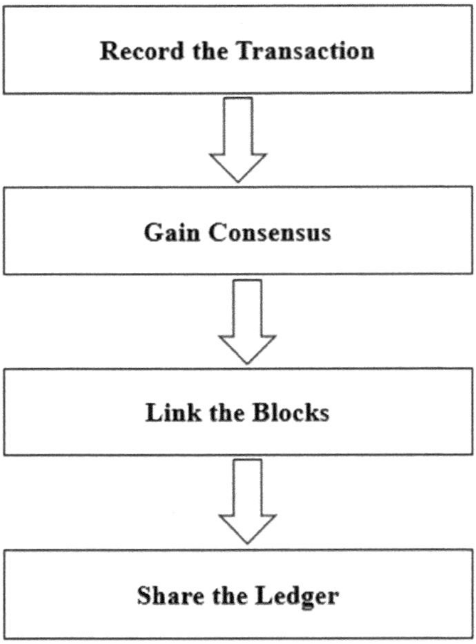

FIGURE 4.4 Blockchain workflow.

4.4 APPLICATIONS OF BLOCKCHAIN

The various applications of Blockchain are as follows [12]:

1. **Energy and Utilities:** Blockchain has the potential to streamline utility billing and settlement processes, enabling peer-to-peer energy trading, tracking the generation of renewable energy sources, and more.
2. **Finance:** Blockchain enables the development of decentralized financial services and platforms, including lending, borrowing, trading, and yield farming.

3. **Media and Entertainment:** Blockchain manages copyright information. This helps safeguard the creators' intellectual property rights by preventing unauthorized use and plagiarism.
4. **Retail:** These businesses use blockchain to track the movement of goods between suppliers and customers [13].

4.5 OPEN-SOURCE PLATFORMS – BLOCKCHAIN

A few of the open-source platforms of Blockchain are explained as follows:

1. **Hyperledger Fabric:** The Hyperledger project of the Linux Foundation has created the open-source, enterprise-grade blockchain architecture known as Hyperledger Fabric. Its goal is to make it possible to create private, permissioned blockchain networks that are appropriate for commercial and enterprise applications. Because of its modular and adaptable architecture, Hyperledger Fabric is suitable for a wide range of businesses and use cases.
2. **Ethereum:** It is an open-source blockchain platform. It enables the creation of Decentralized applications (DApps) and smart contracts.
3. **Corda:** It is an open-source blockchain project used for commercial purposes. With its focus on privacy, scalability, and interoperability, Corda is designed to meet the specific needs of enterprise blockchain applications.
4. **Quorum:** Based on the Ethereum core, Quorum is an open-source enterprise blockchain platform. To make Ethereum suitable for private and confidential transactions, Quorum includes several improvements and tweaks. Quorum has been used in various financial applications, including trade finance, settlement systems, and payment platforms [15].
5. **Solidity:** It is used to develop Ethereum-based applications and smart contracts. It improves the functionality of the Ethereum Virtual Machine and has a syntax that is similar to JavaScript. It is Ethereum's runtime environment. On any blockchain, it is used to create and implement smart contracts. It is an object-oriented language.

6. **Geth:** It is one of the most widely used software that serves as a node for the Ethereum blockchain. Use Geth as a console to type commands and perform operations. If no default parameters are specified, a connection to the Ethereum Mainnet will be established automatically. Save your data on an external hard drive because Geth by default downloads the entire Ethereum blockchain. Platforms like Windows and Linux support Geth. It is built using the Go programming language and provides users with the ability to:
   - Tokens can be moved between addresses;
   - Smart contract creation and execution using the Ethereum virtual machine;
   - Examine the block's past.
7. **Remix:** For tiny contracts, it serves as a compiler. Smart contracts can be deployed, tested, and debugged with its assistance. Its code analyzer helps programmers write the finest possible codes. MetaMask, a browser extension that makes it easier to access distributed Ethereum apps (DApps), connects it to the blockchain. Since it is open-source, anyone may use it for free.
8. **Mist:** It is a graphical user interface for the Ethereum platform. Particularly for the deployment of smart contracts, it is used. It serves as a complete node wallet for storing ether tokens. It requires downloading the full Ethereum blockchain, which takes up more than 1 Terabyte, in order to use it. The program will be updated with the Ethereum blockchain after being downloaded. Once everything has finished synchronizing, you will be given the option to function on either the regular blockchain or the test blockchain. Select an extremely safe password and once it has been generated, it cannot be changed. As a last verification, enter your password each time when send Ether.
9. **Solium:** Every tool's development must take security into consideration. The solidity code needs to be impenetrable and secure. It aids in formatting solidity codes and aids in the correction of any security flaws that may exist. It is a tool made specifically to look for code vulnerabilities. The Solidity Style Guide is not

strictly followed. It supports acceptable code standards that the community has embraced.

10. **Parity:** To publish contracts to the Ethereum network, a development environment must be built up after compilation. Parity comes in as an Ethereum client is needed. It is a safe method of interacting with the blockchain. It uses Rust programming and was created by Gavin Wood, the former CEO of Ethereum. It comes with a web browser pre-installed. It provides customers with access to Ethereum as well as basic wallet capabilities. It is an Ethereum GUI browser because it allows users access to all of the network's functionality.
11. **DApp Board:** It is a reputable Ethereum blockchain explorer. It is a web-based analytics application that serves as an environment for analyzing Ethereum smart contracts.
12. **Truffle:** It is a platform for developing smart contracts. Developers can run automated tests with technologies like Mocha and Chai. It is a tool that aids with migrations and scriptable deployments. It has the following features:
    - Public and Private network management.
    - A console environment for direct contract communication; smart contract compilation, deployment, and linking built in.
13. **Embark:** It is a developer platform that aids in the development of DApps and facilitates the construction of original smart contracts. When a contract is modified, Embark immediately updates the contract and associated DApps. The Embark platform is interacted with using standard web languages such as angular, react, and others. Developers may now manage their contracts across several blockchains thanks to JavaScript's support for test-driven contract creation.
14. **MyEtherWallet:** A paper wallet is the most secure way to keep cryptocurrency. This preserves a substantial portion of your cryptocurrency revenue. Storage is classified into two types:
    i. **Hot Storage:** It is internet-connected, making it easier to spend cryptocurrency, but it is also vulnerable to hackers.

ii. **Cold Storage:** It is used to keep cryptocurrency offline. Even if spending it is difficult, it is harmless from third-party attackers.

A paper wallet, a cold storage technique, includes displaying the private and public keys on a sheet of paper and storing them in a secure area.

4.6 BLOCKCHAIN-BASED DIGITAL TWIN TECHNOLOGY

4.6.1 INTRODUCTION TO BLOCKCHAIN-BASED DIGITAL TWIN

Digital Twin enabled Blockchain to revolutionize the industry landscape through secure data management by Blockchain and predictive maintenance of data by Digital Twin. The advantages of implementing blockchain-based DTs in the industry are as follows [16]:

1. **Reliable Sources:** In industrial operations, IoT-based devices that could be susceptible to cyber threats are used to map physical to digital space. Because the obtained data drives logical decisions, it is critical to establish the dependability of the data source. Blockchain can play an important role in preventing device tampering by collecting information from registered devices.
2. **Immutable and Decentralized Data Transactions of DTs:** It is crucial to protect the DT's platform and data. Blockchain enhances security by introducing distributed ledgers that store DTs data that are hard to alter and uncontrollable by any central authority, as well as by using cryptographic hashing methods. One of the unique features of blockchain is immutability, ensuring that the blockchain ledger remains unchangeable and maintains the unalterable history of DTs transactions. This feature improves the effectiveness and efficiency of the DTs audit process.
3. **Identity and Legitimacy:** All product information is stored in an immutable record through blockchain-enabled DTs. Buyers can also check the transaction records to determine where a specific product was manufactured, who owned it previously, and other details. This simplifies establishing legitimacy and identifying DTs.

4. **Global DT Tracking with High Accuracy:** The blockchain-based generation of DTs can aid in their highly accurate global identification and tracking. Blockchain-based Digital Twin accesses distributed databases safely from several locations and it does not require the participation of a third party. Record-keeping, provenance tracking, and auditability are simplified with blockchain-based DTs due to the ease of access to and tracking of the DTs' information. The usage of blockchain allows one to monitor the evolution of DTs from their genesis to the present. Such tracking features can be utilized to detect fraud in any DT component.
5. **Data Coordination with Trusted DTs and Access Privileges:** Most conventional access control methods are susceptible to hackers. However, blockchain reduces the risks of unauthorized access, which commonly leads to data change. All of the DTs' information, including their policies and regulations, is managed in a distributed manner utilizing blockchain-based access control, allowing each DT to function independently. Furthermore, the secure sockets layer, which manages the data on blockchains, strengthens security in blockchain-based access control. DTs, on the other hand, often use IoT-based technology to interact and cooperate with their actual goods. It reduces security threats.
6. **Transparency and Accountability:** Blockchain imposes transparency and accountability by utilizing encryption and control tools such as smart contracts and chain codes. Each blockchain user is allotted a public address and any authorized user can check these transactions. Furthermore, blockchain technology has user-defined control mechanisms that support accountability enforcement.

4.6.2 BLOCKCHAIN-BASED DIGITAL TWIN MODEL

The blockchain-based DT model includes the following components. They are:

- Physical space;
- Virtual space;

- Services;
- Data wrangler;
- Data synchronization;
- Data fusion;
- Blockchain;
- Composability.

The components of the blockchain-based DT model are depicted in Figure 4.5.

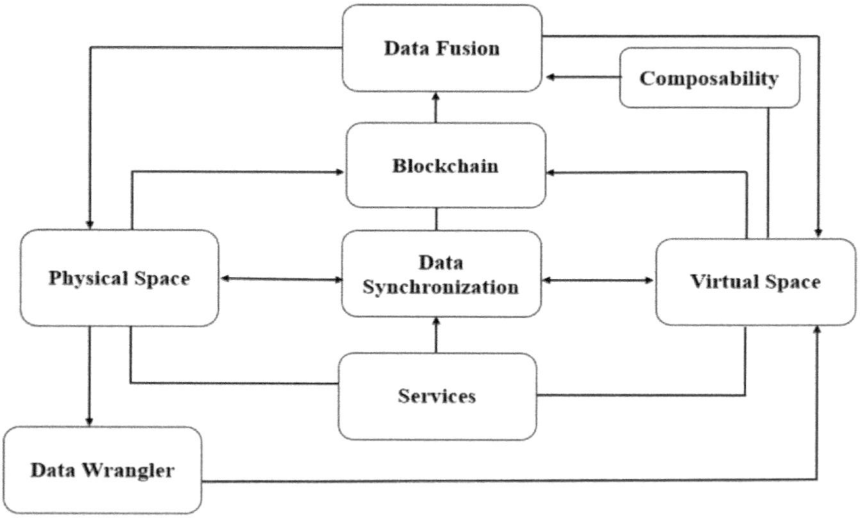

FIGURE 4.5 Blockchain-based digital twin model.

The significance of each component is explained as follows [17]:

1. **Physical Space:** Through asset data, the physical part assists the virtual part.
2. **Virtual Space:** Through decision-making, and simulation, the virtual portion governs the actual portion.
3. **Services:** The service assists in the operation, evolution, and optimization of both virtual and physical areas. The data repository is the key source for data generation in digital twin (DT) and connects all components of the DT. Data-based digital twins (DTs) emphasize product design, manufacturing, production, operational

control, and so on in the industrial realm. DTs, for example, provide more exact predictions, logical judgments, and informed planning by enhancing communication between the design and manufacturing phases. A data-driven DT system performs fault diagnosis, finds the unreliable behavior of the actual object, and provides design principles for maintenance.

4. **Data Wrangler:** The data wrangler is in charge of removing missing or invalid data and converting dissimilar data types into a similar one before entering them into real and virtual environments. During the car manufacturing process, data is received from a variety of physical sources like sensors and actuators, and it is fed as input into physical and virtual domains.
5. **Data Synchronization:** It performs a consistent digital-physical mapping between sensor data gathered from the production unit and inbuilt performance parameters obtained from the storage unit to ensure data consistency.
6. **Data Fusion:** It is the method of binding data from numerous sources to create a consistent interpretation for a certain object. For example, critical information such as order details, supply chain data, simulation data, and consistency checks are already available for usage in the repository before the commencement of the automobile production process. Furthermore, data wrangling and data fusion are critical for obtaining and merging data after the automotive manufacturing process. This information can be used by threat intelligence to research the enemy's domain and alter S&S standards for covert activities.
7. **Blockchain:** It secures data handling for all components. Transactions between nodes serve as the essential building blocks of communication in the blockchain and are maintained in a shared digital ledger as append-only time-stamped records. A consensus technique is used to build a secure trustworthy network. A block of transactions is assembled and appended to the ledger using the consensus algorithm. In the Blockchain framework, the hash value of the block enforces immutability and makes tampering with a block and hence transactions easier to detect.

8. **Composability:** To integrate and communicate with them, the framework takes advantage of composability, which aggregates numerous replicas or DTs.

4.6.3 BLOCKCHAIN-BASED DIGITAL TWIN FRAMEWORK

The diagram of the blockchain-based digital twin framework is shown in Figure 4.6. The conceptual framework consists of three tiers. They are:

- Data;
- Storage; and
- Application.

The process steps of Blockchain based Digital Twin Framework are explained as follows [17]:

- **Step 1:** People, sensors, and nodes are authorized objects on the blockchain
- **Step 2:** The specific parameters are monitored, collected, and processed while being mirrored in the logical space and physical space at the data layer.
- **Step 3:** The gathered data is transferred to the storage layer along with a complete family of data and actions.
- **Step 4:** Models are created at the twinned system and stored at the storage layer using data gathered, domain expertise, and system history.
- **Step 5:** The agile feedback loop is completed when the application layer requests scheduling services or model calibration services after continuously analyzing data to identify manufacturing issues.

Security cameras, electricity monitoring, fire alarms, and other monitoring tools can also be utilized to develop physical security countermeasures. Thus, DTs foresee occurrences and their underlying causes using provenance-enabled trustworthy data, while blockchain offers safe data administration.

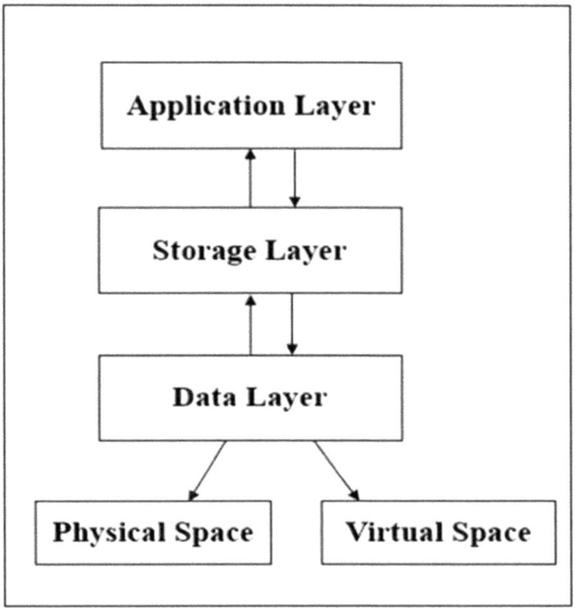

FIGURE 4.6 Blockchain-based digital twin framework.

4.6.4 APPLICATIONS OF BLOCKCHAIN-BASED DIGITAL TWIN ACROSS INDUSTRIES

1. **Utilities:** This sector is utilizing Blockchain and Digital Twin technology to boost productivity. Furthermore, the supply and demand for energy change in response to consumer demands. IoT devices also help with data collection and the development of a transparent, smart energy environment. By emulating the behavior of real meters, digital twins are critical for assessing customer usage. As a result, energy firms can identify user trends and adjust power distribution to their specific requirements. Customers also have real-time access to the data and can tailor usage to their own needs [18].
2. **Logistics:** To begin with, a lack of transparency is producing a slew of issues in the logistics industry. Blockchain and distributed ledger technology (DLT) enhance security and streamline processes in logistical operations. Blockchain is further linked to

IoT sensors. As a result, IoT sensors generate digital twins using blockchain-based data, facilitating the tracking and monitoring of things and products. Furthermore, it provides organizations with full information about a product's progress through transportation. By storing and tracking information about things, blockchain, and digital twins can aid in the resolution of significant logistical difficulties.

3. **Healthcare:** To better its operations, the healthcare industry has long used cutting-edge technology. As a result, innovations in the healthcare sector provide more opportunities to save lives. Digital twins, for example, can duplicate a patient's heart using data from a digital pacemaker. As a result, cardiologists can better understand a patient's condition. It also aids in identifying the best course of action and diagnosis by wirelessly monitoring the patient's heart. Medical and healthcare records for patients can also be maintained on a blockchain and made available to specialists.

4.7 OPEN-SOURCE PLATFORMS FOR BLOCKCHAIN-BASED DIGITAL TWIN

Blockchain-based Digital Twins can be difficult to create and implement, but the appropriate tools make it much easier. Here are some open-source platforms for deploying Blockchain-based Digital Twins.

4.7.1 PROJECT DOVETAIL

Project Dovetail by TIBCO LABS makes it easier to build, visualize, test, and audit smart contracts without requiring considerable programming experience. Low-level code is segregated from logic, and contracts can be easily integrated into a variety of blockchain stacks. Time to market can be sped up, on-chain, and off-chain compute can function more seamlessly together, and there is less risk of technology lock-in. Smart contracts are applications or business logic (written in code) that run within the architecture of a blockchain network.

Project dovetail provides the complete design, development, testing, and deployment of blockchain smart contracts. The modeling abstraction layer enables the creation of smart contracts based on a different kind of models, which may assist to:

- Increase security as reusable, tested code drawn from the model is made possible by the modeling abstraction layer.
- Utilize an extensible abstraction layer to streamline development (flow model).
- Break your code's connection to the underlying blockchain technology.
- Reduce the amount of code needed.
- Boost auditability and visibility.
- Make your industry's modeling customizable.
- Develop a better user interface for designing smart contracts.

Project dovetail is intended to be lightweight, adaptable, and embeddable, enabling the rapid development of blockchain-based solutions.

4.7.1.1 ARCHITECTURE – PROJECT DOVETAIL

Project Dovetail can be conceived of as having four primary sections, from the top down:

1. **CLI:** This tool installs the well-defined smart contracts to the desired blockchain layer.
2. **UI:** Used to create the logic for blockchain smart contracts. Users can use pre-made drag-and-drop activities, add code, or create any required extensions. Data can also be altered and mapped.
3. **Metadata:** Outlines the model(s) on which the blockchain's smart contracts are based.
4. **Interpreter:** Uses the blockchain target's context to carry out the deployed smart contract.

The Project Dovetail solution is made up of these elements, which enable users to create smart contracts that are aimed at different blockchain platforms. The architecture is depicted in Figure 4.7.

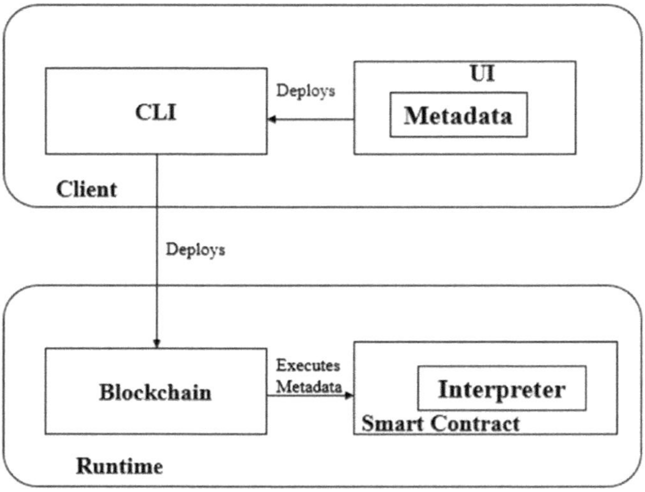

FIGURE 4.7 Architecture – project dovetail.

4.7.2 ETHERTWIN

The EtherTwin prototype is built on a research methodology for communicating Digital Twin data across time [19]. EtherTwin employs a distributed strategy, utilizing the Ethereum blockchain and the distributed hash table (DHT) Swarm, to enable participation from the numerous lifecycle stakeholders without the requirement for trusted third parties (TTPs). Based on Automaton ML (AML) files that identify assets, a Digital Twin can be created and shared with the parties engaged in the twin's lifetime. The EtherTwin prototype allows:

- Create digital twins;
- Sharing each twin, uploading paperwork, and modifying specs for each twin (versioning);
- Create sensor data feeds;
- Manage user access using asset attributes and lifecycle roles;
- Make a list of the asset's components (specification parsing).

4.7.3 SMARTTWIN

SmartTwin is a Blockchain-enabled Digital Twin software architectural structure [20]. The key elements of Digital Twin architecture involve the mixture of IoT (Internet of Things) devices as direct actors on the blockchain system and exploiting Blockchain's cryptographic capabilities to ensure data security. Enterprise Digital Twin denotes the interaction of a person, equipment, and analytical software to create a paradigm relationship between the physical and digital worlds. Controls and enhances production cycle processes, simplifying operating activity, lowering failure frequency, and boosting equipment service life.

4.7.4 REMIX IDE

Smart contracts can be created using Remix IDE, which can then be compiled and deployed [21]. Ethereum is the favored platform for smart contract developers since it is the most extensively used platform for building smart contracts and includes the Solidity programming language for scripting smart contract code. In contrast, a specific piece of code is only useful if it can be compiled into machine-readable language. The Remix IDE is the preferred compiler for Ethereum smart contracts.

4.7.5 TIBCO CLOUD™ LIVE APPS

With TIBCO CloudTM Live Apps, a low-code application development platform, businesses can accomplish all of this and more [22]. Although no graphical design-time tooling is now available on blockchain platforms, the visual modeling in Live Apps simplifies the implementation of blockchain solutions for existing systems or new systems. Furthermore, Live Apps is a fantastic addition to blockchain technology since it enables the quick, easy, and iterative building of process applications.

4.7.6 WATSON IOT PLATFORM

Watson IoT Platform was developed by IBM. It is an IoT data tool that gathers data from IoT gadgets and is used for managing large-scale real-time systems. This tool contains extra features like cloud services, blockchain services, data analytics, and edge computing [23].

4.7.7 SOLIDITY

Solidity is the most often utilized coding language for building Ethereum smart contracts [24]. It is largely used for the creation of smart contracts on Blockchain platforms. Smart contracts are programs that run on Blockchain, use business logic, and produce a sequence of transaction records. It executes the code in an Ethereum virtual machine (EVM) [25].

4.8 CONCLUSION

The bulk of digital twin information is centralized, which does not support data management security, integrity maintenance, or the trustworthiness and precision of time-consuming procedures appropriately. To address these constraints, blockchain must be coupled with digital twins. Blockchain technology, in conjunction with Digital Twins, provides important information for decision-making while also assuring the storage of data in a decentralized manner, sharing of data, peer-to-peer communication, and scalability. Combining Blockchain with Digital Twins has the potential to drastically improve a variety of industries. As a result, the two technologies improve the security and efficiency of industries that use them. They also assist businesses in widening their horizons.

This chapter discusses the overview of both Blockchain and Digital Twin technologies in terms of features, workflow, applications, and open-source frameworks. Open-source frameworks of Blockchain such as Hyperledger Fabric, Ethereum, Corda, Quorum, Solidity, Geth, Remix, Mist, Solium, Truffle, Parity, DApp Board, Embark, MyEtherWallet, etc., have been discussed. This chapter also discusses a few open-source tools of Digital Twin such as Ansys Digital Twin Builder, Ditto, Kafka, AWS Digital Twin, AWS IoT TwinMaker, Twinbase, etc.

It emphasizes the benefits of a Blockchain-based Digital Twin model and explains the deep understanding of the building blocks of a Blockchain-based Digital Twin model. It emphasizes the conceptual framework of a Blockchain-based Digital Twin. It explains the significance of a few open-source platforms like Dovetail, EtherTwin, SmartTwin, Remix, Solidity, TIBCO Cloud™ Live Apps, Watson IoT Platform, etc., for implementing a blockchain-based digital twin.

The integration of Blockchain and Digital Twins will provide a safe, cost-effective, and scalable solution that combines digital identification and data tracking via blockchain traceability. Blockchain reduces the cost of Digital Twins, allowing them to be built for a wider range of sectors.

KEYWORDS

- **blockchain**
- **data tracking**
- **digital identification**
- **digital twins**
- **open-source platforms**
- **open-source tools**

REFERENCES

1. Celik, Y., Petri, I., & Rezgui, Y. (2021). Leveraging BIM and blockchain for digital twins. *International Conference on Engineering, Technology, and Innovation (ICE/ITMC)–IEEE, 18*, 1–10.
2. Barricelli, B. R., Casiraghi, E., & Fogli, E. (2019). A survey on Digital Twin: Definitions, characteristics, applications, and design implications. *IEEE Access, 7*, 167653–167671.
3. Aruchamy, P., Sabeena, G., & Sowndarya, K. (2023). An artificial intelligence approach for energy-aware intrusion detection and secure routing in the internet of things-enabled wireless sensor networks. *Concurrency and Computation Practice and Experience*, 1–28.
4. Segovia, M., & Garcia, J. (2022). Design, modeling, and implementation of digital twins. *Sensors–MDPI, 22*, 1–30.
5. Singh, M., Srivastava, R., Fuenmayor, E., Kuts, V., & Qiao, Y. (2022). Applications of Digital Twin across Industries: A Review. *Applied Sciences, 12*, 1–28.
6. Hinduja, H., & Shreya, K. (2020). Industry 4.0 Digital Twin and its Industrial Applications. *International Journal of Science Engineering and Technology, 8*, 1–7.
7. Botkina, D., Hedlind, M., Olsson, B., & Henser, J. (2018). Digital Twin of a Cutting Tool. *51st CIRP Conference on Manufacturing Systems–Elsevier, 72*, 215–218.
8. Autiosalo, J., Siegel, J., & Tammi, K. (2021). Twinbase: Open-source server software for the Digital Twin Web. *IEEE Access, 4*, 1–20.

9. Bhaskar, K. B., Prasanth, A., & Saranya, P. (2022). An energy-efficient blockchain approach for secure communication in IoT-enabled electric vehicles. *International Journal of Communication Systems, 35*, 1–20.
10. Akash, S. S., & Ferdous, M. S. (2022). A Blockchain-based system for healthcare digital twin. *IEEE Access, 10*, 50523–50547.
11. Aste, T., Tasca, P., & Matteo, T. D. (2017). Blockchain technologies: The foreseeable impact on society and industry. *IEEE Computer, 50*, 18–28.
12. Moez, K., Meryem, A., Alaeddine, M., & Mutiq, A. (2022). Blockchain for modern applications: A survey. *Sensors–MDPI, 22*, 1–27.
13. Capocasale, V., Danilo, G., & Perboli, G. (2023). Comparative analysis of permissioned blockchain frameworks for industrial application. *Blockchain: Research and Applications–Elsevier, 3*, 1–13.
14. Agrawal, K., Aggarwal, M., Tanwar, S., & Sharma, G. (2022). An extensive blockchain-based applications survey: Tools, frameworks, opportunities, challenges, and solutions. *IEEE Access, 10*, 116858–116906.
15. Espel, T., Katz, L., & Robin, G. (2017). Proposal for protocol on a quorum blockchain with zero knowledge. *International Association for Cryptologic Research*, 1–22.
16. Sahal, R., Alsamhi, S. H., Brown, K. N., O'Shea, D., & Alouf, B. (2022). Blockchain-based digital twins' collaboration for smart pandemic alerting: Decentralized COVID-19 pandemic alerting use case. *Computational Intelligence and Neuroscience, 2022*, 1–14.
17. Suhail, S., Hussain, R., Jurdak, R., Oracevic, A., Salah, K., & Matulevicius, U. R. (2021). Blockchain-based digital twins: Research trends, issues, and future challenges. *ACM Computer Survey, 54*, 1–34.
18. Raj, P. (2021). Empowering digital twins with blockchains. *Public Health Emergency Collection–Elsevier, 121*, 267–283.
19. Putz, B., Dietz, M., Empl, P., & Pernul, G. (2021). EtherTwin: Blockchain-based Secure Digital Twin Information Management. *Information and Processing Management, 58*, 1–23.
20. Pincheira, M., Prieto, J., Luis, F., Martinez, B., & Ferretti, S. (2023). SmartTwin: A Blockchain-Based Software Framework for Digital Twins Using IoT. *Lecture Notes in Networks and Systems Book Series–Springer*, Chap. 7.
21. Khandelwal, P., Johari, R., Gaur, V., & Vashisth, D. (2021). Blockchain Technology-based Smart Contract Agreement on REMIX IDE. *International Conference on Signal Processing and Integrated Networks–IEEE*, 938–942.
22. TIBCO. (2023). TIBCO cloud live apps. Retrieved from: https://www.tibco.com/connected/tibco-cloud-live-apps (accessed on 25 July 2024).
23. Sangeetha, A. S., Shunmugan, S., & Murugan, G. (2020). Blockchain for IoT enabled supply chain management–A systematic review. *International Conference on I-SMAC–IEEE, 7*, 48–52.
24. Wohrer, M., & Zdun, U. (2018). Smart contracts: Security patterns in the Ethereum ecosystem and solidity. *International Workshop on Blockchain Oriented Software Engineering (IWBOSE)–IEEE, 10204*, 1–48.
25. Zheng, G., Gao, L., Huang, L., & Guan, J. (2021). Solidity basics. *Ethereum Smart Contract Development in Solidity–Springer*, 49–53.

CHAPTER 5

Modeling and Simulation for Digital Twins of Blockchain Technology

U. HEMAMALINI,[1] A. BHARATHI,[1] A. PRASANTH,[2] P. MANJULA,[3] and K. K. DEVI SOWNDARYA[4]

[1]*Assistant Professor and Research Scholar, Department of Information Technology, School of Computing Sciences, VISTAS Pallavaram, Chennai, Tamil Nadu, India*

[2]*Associate Professor, Department of Computer Science and Engineering, Vel Tech Rangarajan Dr. Sagunthala R&D Institute of Science and Technology, Chennai, India*

[3]*Associate Professor, Department of Computer Science and Engineering, Saveetha School of Engineering, Chennai, Tamil Nadu, India*

[4]*Assistant Professor, Department of Electronics and Communication Engineering, DMI College of Engineering, Chennai, Tamil Nadu, India*

ABSTRACT

A virtual twin is an exact representation or counterpart of any real system or process. A digital twin is sometimes known as a "virtual twin." The simulation of the digital twin can infer how an object, product, or system will function in the future after all of the essential structural modifications have been made. This is entirely feasible because a simulation based on a digital twin can recognize an object's material, distinguish moving parts from static ones, and, finally, the

Blockchain-Based Digital Twins: Research Trends and Challenges.
A. Prasanth, Vandana Sharma, Rajesh Kumar Dhanaraj, Pethuru Raj (Eds.)

use the simulated product can be programmed into the digital platform to accurately represent its functioning. The manufacturing sector has already adopted the technology of digital twins quite some time ago in order to bring their processes, outputs, and products into the current era. Since then, the unique contributions made by the digital twin idea have substantially increased thanks to flawless synchronization with a variety of pioneering technologies. Those technologies can optimize a wide range of processes with the help of digital twins, such as process excellence, information discovery, and distribution at the appropriate moment. The potential for digital twins has recently seen a remarkable expansion because of the development of blockchain technology. The following chapter discusses the various modeling and simulation techniques utilized for digital twins. A detailed description of these techniques assists the researchers in establishing several real-time applications related to digital twins.

5.1 INTRODUCTION

It is to be anticipated that, as time goes on, the requirements of enterprises and the expectations of individuals will continue to morph and grow. The requirements are being easily and elegantly met thanks to the proliferation of groundbreaking technology, both individually and jointly [1]. Because of the proliferation of Internet of Things devices in the surroundings in which we live and work, there is a growing demand to endow these devices with the necessary intelligence so that they can demonstrate adaptive behavior in their actions and reactions [2]. The transformation of connected devices into intelligent ones is a requirement.

There are a number of technical options, including Machine Learning and Deep Learning algorithms, and integrated data analytics systems that help speed up the transition process. That is, the data that has been obtained has been put through a number of distinct investigations that are more in-depth and definitive in order to draw out information that is helpful and insights that are actionable. Many people have proposed a variety of strategies and approaches that can be used to endow Internet of Things devices with intelligent capabilities. Creating digital twins for Internet of Things devices is one of the options available to you.

The purpose of digital twins is to continuously collect data that adds value to their interrelated physical twin and then analyze that data to generate actionable insights out of mountains of data. To achieve and obtain the value that has been promised for digital artifacts, however, there is a requirement for the highest possible level of security and transparency due to the rapid growth of digital assets and data everywhere. Therefore, the unique integration of digital twin and blockchain technologies is being affirmed to provide an effective method to alleviate any security holes and boost the confidence of users in the utilization of digital assets for their day-to-day transactions [3]. This assertion and presentation are being made in light of the fact that these technologies have recently been brought together for the first time. This is being done in order to make the most of the benefits that each of these technologies has to offer.

5.1.1 DIGITAL TWIN

In the view of humans, the term "digital twin" refers to the entirety of the data footprint that was produced as a result of the characteristics, interactions, and online presence. When the aspects that enable decisions are joined with one another in a serious way and when the aggregated outputs are processed by sophisticated and real-time artificial intelligence and data analytics technologies, it is feasible to precisely anticipate our health status and remaining tendencies [4].

1. **Attributes:** These are a person's name, age, gender, address, ideology, language, culture, religion, ethnicity, education level, and salary, among other things. All of these data points are meticulously collected and saved in digital format. As a result, the creation of digital twins for humans is feasible.
2. **Interactions:** This is all about how we interact with the outside world on a regular basis. When trying to exactly estimate our state of health, the specifics of where and when we are traveling, as well as the restaurants we frequent, are really helpful. In convergence of unique characteristics, it can forecast the patient's state of health and provide treatment recommendations.
3. **Online Presence:** This encompasses a variety of professional and social media apps that convey individuality. As a result, human

digital twins are becoming increasingly popular. A digital twin of a human being can help determine whether or not that person is at risk of contracting a disease that could endanger their life. Additionally, it can help provide insights on how to change one's lifestyle in order to be healthy and live for a longer period of time.

Digital copies of physical twins are currently undergoing painstaking construction. Utilizing their education, experience, and competence in computer algorithms, data science, statistics, mathematics, and other related fields, software specialists have begun developing futuristic digital twins. As a result, digital twins provide a significant contribution to the process of including new features in pre-existing physical systems by recognizing and expressing the benefits and drawbacks of implementing the features.

There is a further adjustment that is known as the "predictive twin." This forecast is generated by analyzing previously collected data from other devices. The intrinsic potential of predictive twins to replicate breakdowns and other events that require prompt attention is one of their defining characteristics [5]. Microsoft has introduced a new word, which the company calls "process digital twin." There are some sophisticated manufacturing scenarios that can be illuminated with the help of process twins that cannot be provided by product digital twins.

5.1.2 BLOCKCHAIN

For the digital twin paradigm to achieve the level of success that was envisioned, data is of the utmost importance. In a linked world, new data streams would typically consist of the attributes of the participating physical items, the data pertaining to their interactions with other physical things, as well as their future states. Both the physical and digital worlds participate in the data exchange that takes place. Digital twin instances, also known as digital twins, are simulations of real-world entities and activities [6]. On the other hand, digital twin aggregates, also known as digital twins, are collections of digital twins that work together to accomplish more and create higher-quality results.

A violation of a digital twin will result in the compromise of one product as a result of this combination and collaboration. In the event of a breach of the digital twin, the involved items will be put at risk. The nascent blockchain technology has the inherent capacity to overcome safety and security gaps, which are necessary in order to get digital twin projects back online. Because blockchain technology possesses the characteristics of decentralization and immutability, digital twin projects that make intelligent use of blockchain can innovate more effectively and more quickly than they could otherwise [7].

The cryptographic properties of the blockchain make it possible to transfer data in a safe and reliable manner [8]. Multiple parties, including users, collaborators, and stakeholders, can be gathered under a single roof by utilizing one or more of the consensus mechanisms that are built into the blockchain technology. The stakeholders have evolved into the gatekeepers, who are responsible for enforcing transparency and accountability provided by blockchain technology and are used to empower digital twins. As a result, sensitive information about both the company and its customers can be shared without worry.

The Internet of Things, digital twins, artificial intelligence, and other cutting-edge technologies are being smartly integrated and are empowering blockchain technology [9]. The astute synchronization of these technologies with blockchain, on the other hand, is providing them with a boost that is both succulent and significantly empowering. It is common knowledge that one of the most significant challenges facing businesses all around the world in the modern era is the expert forging of their products. Thanks to the development of modern technology and tools, fraudsters now have the ability to quickly produce duplicates of a wide variety of high-priced products, which they can then sell to unwary customers at a reduced cost. This precarious circumstance not only tarnished the reputation of the brand but also resulted in significant financial losses [10].

When combined, digital twin technology and blockchain have the potential to deliver an effective answer to the problem of combating certain types of fraud. Together, they provide reassurance that the products are legitimate. The digitalization of tangible artifacts is made easier with the help of digital twins. Blockchain, with its capacity for decentralization, will bring about the transparency that is so desperately required. Any digital data that is stored will benefit from this increased level of protection.

5.1.3 BLOCKCHAIN FOR DIGITAL TWIN

The introduction of blockchain technology has recast the trajectory of development for digital twins. At a macro-level, blockchain technology makes it possible to store data in a distributed manner without a central authority figure having access to all of it [11]. New transactions may be added, but the ones that are already there cannot be altered or corrupted in any way. The integrity of the data and the transparency of the process are both ensured by this. The creation of a digital twin on a blockchain would make it possible to record in an unchangeable format any and all data relating to a physical product. Additionally, the transaction records associated with the product can be retained.

A digital twin that is constructed using conventional technology is required to have a centralized intermediary in order to store digital data and do data analytics [12]. The idea of digital twins will almost certainly go through some kind of fundamental adjustment as a result of the rapid development of blockchain technology. The immutability and security that are both assured by blockchain technology will propel digital twins to the next level. Evidently, blockchain is currently being promoted as a technology that may prevent counterfeiting as well as theft. If companies all around the world put this into practice with the due care and attention it requires, then the problems caused by forgery and fabrication will be much easier to solve.

The confidence that individuals have in a variety of products receives a considerable boost [13]. Instantly, purchasers have access to all the important information regarding any product. Everyone would have access to in-depth information about each product, and the idea of unconfirmed ownership would become a thing of the past as a result. This invigorating combination is making its way into every industrial vertical, from the manufacturing sector to healthcare, advertising, and the logistics industry.

5.2 BENEFITS OF USING BLOCKCHAIN FOR DIGITAL TWIN

There are numerous benefits of using blockchain for Digital twins as depicted in Figure 5.1.

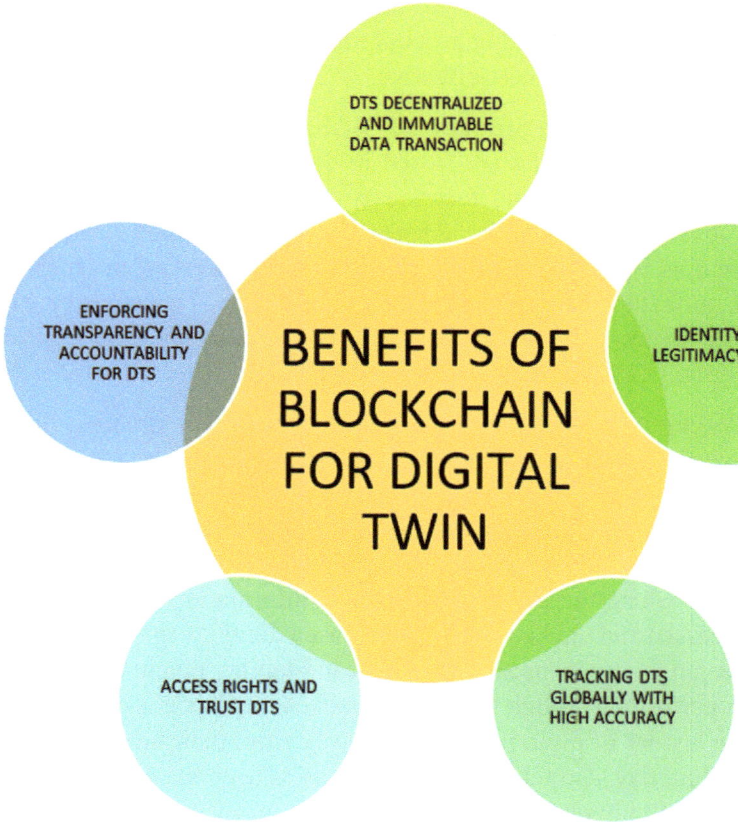

FIGURE 5.1 Benefits of blockchain-based digital twins.

1. **DTS Decentralized and Immutable Data Transactions:** It is essential that both the platform and the data of digital twins be protected. The use of cryptographic hashing techniques and the storage of digital twin data on distributed ledgers, which cannot be readily altered and are not under the jurisdiction of any centralized authority, contribute to the increased security provided by block-chain technology. One of the characteristics that sets blockchain apart from other distributed ledger technologies is its immutability, which assures that the distributed ledger can never be altered and maintains an unchangeable record of all digital twin transactions. This feature not only adds a degree of trust that has never been

seen before and ensures the integrity of digital twin data, but it also makes the auditing process for digital twins a method that is both efficient and cost-effective.

2. **Identity and Legitimacy of DTS:** Immutable copies of all product-related data are stored in digital twins that are empowered with blockchain technology. In addition, records of their transactions are kept, which enables buyers to learn information such as the location of the factory where a certain product was manufactured, its previous owners, and so on. This contributes to the creation of evidence that verifies the legitimacy and identity of digital twins.
3. **Tracking DTS Globally with High Accuracy:** Creating digital twins on a blockchain platform can make it easier to identify and keep track of them around the globe with more precision. Blockchain enables decentralized parties to securely access distributed databases from multiple locations without the need for a third party. Since the distributed ledger's information can be quickly accessed and tracked, blockchain-based digital twins make record-keeping, provenance tracking, and auditability much simpler tasks to perform. Through the use of blockchain, it is possible to track the history of digital twins all the way from their inception to the present day. A tracking method that is as accurate as this one can be used to investigate and locate instances of fraud in any aspect of digital twins.
4. **Access Rights and Trusted DTS:** The majority of traditional access control measures are susceptible to being attacked by malicious cyber attackers. Blockchain, on the other hand, reduces the hazards of illegal access, which is the most common cause of tampered data. Access control that is based on a blockchain is able to manage all of the digital twins' information in a distributed manner, including policies and rules. This allows all of the digital twins to act independently. In addition, blockchain-based access control improves safety by using secure socket layer certificates, which can be found on each digital twin and whose data can be maintained on blockchains. These certificates are stored on the distributed ledgers. On the other hand, digital twins typically utilize technologies based on the Internet of Things in order to

coordinate and interact with their actual objects. These kinds of coordinated and communicated data are susceptible to security risks, which can be mitigated through the use of blockchain technology.

5. **Enforcing Transparency** and Accountability for DTS: Through the use of encryption and other control measures, the blockchain ensures openness and accountability (e.g., smart contracts and chain codes). Every blockchain participant is given a public address that is completely transparent, allowing any authorized user to view the holdings and transactions of other participants. Without a shadow of a doubt, this guarantees an unprecedented level of openness across the digital twins. In addition to this, blockchain technology relies on user-defined control mechanisms, which aid in the implementation of accountability measures.

5.3 DIGITAL TWIN DESIGN

In order to complete the first stage of the process of developing a digital twin, it is important to model the static characteristics of the system. The purpose of this design is to determine the parameters and requirements of the system, including its functions and the breakdown of those functions. It is important that the modeling of data exchanges and communication, as well as the structures and reasonable layout of the system, as well as the conceptual prerequisites for putting the remedy into practice, which include both the software and hardware components, be taken into consideration.

In most cases, the Virtual Object (VO) is developed so that it can function as a reflection of the Physical Object (PO). As a consequence of this, it is possible for both components to coexist for the duration of the digital twin's existence. After that, the primary focus of the early stage will be on the duties involved in connecting the VO and the PO, assuming that the PO is actually present at the moment the design is being carried out. When the VO is still in the design stage, there is a possibility that its PO element will not even appear at all. In this scenario, a prototype PO can be developed in order to direct the development of the VO. In addition, certain design decisions can be evaluated by combining the model with the VO. Each VO will, in the end, reflect and replicate the PO or its model. This will be the case regardless of whether or not the PO existed when the concept was conceived. It may be possible to anticipate certain problems in the product by using simulation, particularly in situations where the product is subjected to strain or is used in hazardous conditions [14]. The digital twin's ongoing synchronization with both the manufacturing systems and its development are significant qualities; for instance, updates in cabling or modifications to the physical or fixed point should be taken into account in the VO. As a result of this, a solution that is developed needs to be adaptable and able to take into account any recent shifts in engineering models or operational procedures.

5.3.1 SPECIFICATIONS OF THE SYSTEM

The goals and functional needs of the digital twin were originally established by the function model. This determines, for example, whether the digital twin will help with safeguarding, regulating, anticipating, or optimizing the processing. The next thing that needs to be done is to figure out which parts or processes within it will be part of the digital twin. In order to accomplish this, we need to create a plan for the process that takes into account both the procedure and the digital twin features, select the devices to be used, select the modeling aspect that will be used (such as mechanical systems, 3-dimensional areas, electrical, or any others), and determine the connection between the two parts, i.e., how well the parts will communicate with one another. In addition to this, it is necessary for us to determine which digital twin demands are not operational [15].

5.3.2 FUNCTIONAL REQUIREMENTS

This phase will outline the goal as well as the functional requirements in order to develop a digital twin's specification. Additionally, it details the data that we are required to retrieve and explain in compliance with the PO. Digital twin specifications must be defined with precision, and as much data as is humanly possible must be gathered in order to design the system. It is absolutely necessary for the process of system modeling to have a fruitful requirement analysis. Digital twins have been used for a wide variety of purposes, including the optimization of processes, security monitoring, forecasting, education, and training. A digital twin will typically integrate one or more of these capabilities into a single offering.

1. **Optimization:** It is possible to study the complete output of data generated by the digital twin using sophisticated data analysis tools in order to obtain reliable information. Therefore, a digital twin has the potential to improve the functionality and efficiency of a system, as well as the ability to minimize costs or risks, and has gained wide attention. The digital twin is able to run simultaneously with a CPS, which allows them to conduct activities such as analyzing the level of safety and security provided by the system even when it is being used with a CPS in order to perform tasks such as analyzing the level of safety and security provided by the system while it is being used. In their study, Willard et al. [16] conducted a comprehensive review of the deployment of a digital twin in the context of a cyber safety program in CPS. CPSs are composed of two levels that are interconnected, and those levels are the core network layer and the digital layer. Both components are shielded from potential hazards so that the system can continue to work normally.

   The primary applications are always to enhance compliance, increase safety layout, discover mistakes, conduct pentests, and provide security training. The data within the digital twin, which is derived from the data in the V0, enables both the detection of attacks and the restoration of the system's state. Another advantage of a digital twin is that it can be used to build a detailed evaluation, which ultimately results in a more accurate risk analysis. This can be done by applying a digital twin. It is conceivable, for instance, to cause disruptions in the system to test out new situations and analyze how the system reacts to them.

2. **Tracking and Forecasting:** Detectors are maintaining round-the-clock vigilance over the PO, gathering and storing data in a nonstop manner and doing so in a continuous fashion. The purpose of this endeavor is to put the information that was sensed to use so that it may be processed and used to create predictions about the future. This helps to organize the working teams, govern the PO, and achieve better results by establishing synergies and properly managing their time, which all contribute to an increase in overall productivity. One other use for digital twins is in the process known as "development improvement," which evaluates

and enhances the operational capabilities of a new module through the application of design thinking. A functional analyst is required in order to identify problems with traditional product designs and devise solutions to such problems. A digital twin makes it possible to do in-depth data analysis, which, in turn, paves the way for more accurate design decisions and larger overall product enhancements.

3. **User Training:** Digital twins provide an environment in which operators can receive training at a minimal cost and in an environment with a low risk of injury. By utilizing a dynamic environment, they are able to improve their control over the actual operation of the system, which is especially beneficial during maintenance, adverse, or emergency circumstances. Operational situations [17]. In addition to this, it improves the technicians' ability to make decisions and decreases the repercussions of actions that are carried out incorrectly or in an imprecise manner. The transition to Education 4.0, the formation of a more highly qualified labor force, and the acceleration of judgment in the manufacturing division are all helped along by digital twins. To achieve this goal, it is important to implement a user-system interface in order to connect with the digital twin and integrate these skills with virtual reality.

5.4 MODELLING WITH DIGITAL TWINS

A digital twin's most important part is its collection of virtual models. The development of rising simulated representations that accurately replicate the structure, physical qualities, behaviors, and constraints of the PO is, as a result, the phase in the design of a digital twin that is considered to be the most important. The actual statistics from the real world that are stored in a digital twin might also be required to reproduce the processing and track its development, as well as to update the virtual representations.

5.4.1 MODELLING A COMPONENT

In order to construct a digital twin, one has to develop an abstract prototype that represents the physical world. In this section, we will take a look at the modeling process that was utilized in order to derive the VO of POs. Figure 5.2 illustrates the procedure of the data representation.

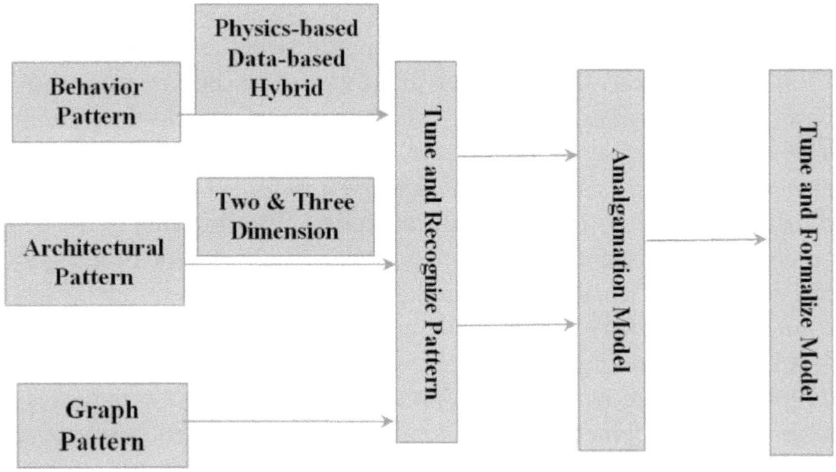

FIGURE 5.2 Data representation procedure.

5.4.1.1 COGNITIVE MODELS

These kinds of models detail how the system will function in accordance with the physical process that the PO is in charge of. In order to comprehend how different physical phenomena, such as power, speed, angular position, and angular speed, communicate with one another. A numerical or computer-based depiction of the relationship between the factors that are relevant to the circumstance is what is meant by the term "models." Control strategies and database strategies are the two primary ways that basic behavioral models are created. It does offer tools for reasoning about system behavior, which can be used to generalize the models to apply to comparable problems in addition to bound mistakes.

The models that are based on data behave more like black boxes and do not provide these advantages. These methods have the advantage of being more adaptable and being able to take into account recent findings from experiments. They may also be better suited for complicated systems, such as 5G networks, in which the application of physics models is either impossible or where it is impossible to comprehend the physics. There is no way to estimate the errors, and the dataset may have introduced bias into the models that were developed using the data [18].

5.4.1.2 CONTROL MODELS

These are models that are based on physics, and they compare the results of simulations with data that has already been gathered and is expressed by statistical equations [19, 20]. The digital twins that are designed for CPS have power as their primary focus and overriding purpose. For the purpose of ensuring continuity, data is gathered in real-time from the underlying processes using sensors. The creation of a behavior model that is founded on control theory can be accomplished through the use of transfer functions or state-space modeling. The first step in the design process uses differential equations to depict the behavior of a regulated physical system. This method simulates the results of the model using the method described above. The mathematical expressions that are connected to the process can then be used to construct the transient response that is linked to the process. It is possible to derive transmission purposes and state-space modeling from one another because these two representations are equivalent to one another [21]. Research was conducted on how to replicate its programmed modes of equipment to achieve their goal of obtaining a simulated version of the CPS while it is operating [22]. In this way, the stimulus is responsible for initiating state changes. This method relies on the system by using the findings of the system's inputs and outcomes in conjunction with a numerical model that represents the system's dynamics [24].

5.4.1.3 DATA-DEPENDENT REPRESENTATIONS

Data-dependent prototypes are prototypes that are based on statistical formats, so this keeps all the factors determining reality at the selected degree of abstraction. The data that is provided by the PO can be used in conjunction with open-source artificial intelligence tools such as TensorFlow [25], PyTorch [26], or OpenCV [27] to generate the VO. In order to produce this type of model, a four-step process must be developed. This process must include the generation of data, the collection of information, the pre-processing of information, and the analysis of facts using AI algorithms. The sensors that collect data from the PO are critically important to the process of creating data. In addition, there are other data-gathering platforms, such as Apache Kafka [28], providing real-time data collection for the data that is gathered. Later, the device will have access to a massive volume of information.

The pre-processing is necessary in order to guarantee that the data are of high quality and comprehensive. As part of the procedure of data engineering, "extracting the features" involves repairing or erasing erroneous and corrupt information. In addition, the integration, transformation, and enrichment of data are all part of the process of data engineering. Apache Spark [29] is an efficient tool for data processing that is memory-based. The processes have the effect of increasing the precision of the data while simultaneously reducing the amount of money spent on computations.

5.4.1.3.1 Models for Hybrid Control-Data

The goal of the fusion prototypes is to bring together the most beneficial aspects of both design approaches [30]. Applications that run in real-time can greatly benefit from the excellent data adaptation that machine learning models provide [31]. For instance, Chakraborty et al. [31] have suggested combining science-based models with learning algorithms in order to develop a digital twin that is capable of predicting the deterioration of structural components. With the help of this method, a rapid digital twin can be linked to the PO in order to provide real-time engineering decision support utilizing an interpretable model (physics-based). The digital twin may be continually improved over time by making use of historical data to educate machine learning models.

Chakraborty et al. [32] have proposed a fusion control scheme for dynamic linear systems with one level of freedom that evolves across two different implementation time scales. This model is applicable to systems with two distinct operational time scales. A physics-based model is used for the method's data handling and reply to prediction, and a machine learning approach is used for the method's analysis of the temporal evolution of the system parameters. Both models are used in conjunction with the method.

5.4.1.4 STATISTICAL MODEL

The statistical model defines the linkages and assemblages of relations between the components that carry out the roles and behaviors. These structures are responsible for carrying out functions and behaviors. In most cases, the structural model will consist of defining the topology, planning the layout, and designing the buffers [33]. Cracking, corrosion, and deformation are some of the phenomena that can be studied using a physical model. The kinetics, reasoning, and connections of the actual system are all replicated in a geometric model [34]. The curves and surfaces that make up the final representation of the 3D models can be generated in one of two ways: either by performing a 3D scan of the object or by employing equations within specialized software [35].

Image-based methods can serve as a suitable alternative to scanning technologies in cases where geometry measurements are required. Additionally, they can be supplemented by information about the object's internal structure, which can be acquired using traditional inspection techniques, thermography techniques, or radar methods, all of which enable a more in-depth examination of a physical structure. This information can be gleaned by using the aforementioned techniques. Anbalagan et al. [36] have described the process of developing digital geometry models through the use of computer-aided design.

5.5 TECHNIQUE FOR THE INTEGRATION OF MODELS

Digital twins will be the integration of complicated models that are connected to one another, producing a view of the system that is more all-encompassing. The integration needs to take a number of different considerations into account, including the following:

1. **How do different parts of a system interact with one another to form more complicated ones?:** This suggests that models work together in order to accurately represent the actions of POs. It is important to keep in mind that the conclusions reached by other independent models can be altered or made incorrect based on the decisions made by some models. As a consequence of this, it is possible for there to be outcomes that are inaccurate or contradictory if the models do not communicate information to one another and do not act in a coordinated manner.
2. **How does the digital twin connect with the environment, specifically how the decisions made by the digital twin have an effect, either directly or indirectly, on the physical processes that are taking place?:** It might be challenging to determine the

bodily impact right off the bat. The selections should be transmitted by the digital components that use the physics principles of the PO so that the effects of these choices and any potential inconsistencies may be assessed. This will allow the digital components to determine whether or not there are any discrepancies.

It is possible for errors to occur in large and varied projects if the relationships and dependencies involved are not adequately described. This indicates that even if each model is exact when viewed in isolation, the exactness of the entire system may depend on how the models interact with one another. This is the case even if each model is accurate when viewed in isolation. Because of all of these factors, the step of model integration in the digital twin development process is one that is both difficult and essential.

5.6 DIGITAL TWIN DEVELOPMENT

The second stage of the procedure consists of putting the digital twin into action. This section looks at the standards, procedures, and protocols that were utilized in the process of creating the PO and synchronizing it with its VO.

5.6.1 PROTOCOLS OF COMMUNICATION AND SYNCHRONIZATION OF DATA

The network is absolutely necessary for the interaction that must take place between the PO and its VO. The link makes it possible to not only collect data in real-time but also exchange instructions with one another. Because there is data exchange in both directions between the PO and its VO, the information that is gathered starting from the PO is used to bring the positions that are held in the digital model up to date. As a consequence of this, the physical system receives input based on the judgments and decisions made within the virtual representation. PO-to-VO interaction typically needs to go through these 3 stages in order to be successful.

i. Gather the pertinent data, which may include precise observations of the real objects;

ii. Analyze the information in a manner that is compatible with the degree of abstraction;
iii. Use the data.

There is more than one way to connect sensors and gateways, and the method that is used will depend on the kind of system that is being used. Numerous commercial solutions have been developed with the goal of offering services and solutions for Internet of Things devices and enabling data processing and transfer. This mode of interaction can also be utilized to relay just important incidents rather than providing information on a more routine basis.

Platforms such as Apache Kafka and protocols such as MQTT are able to handle this type of communication between users. The approach makes it possible to exchange information in both directions using APIs and files. This would be accomplished by using the framework. It places an emphasis on the remote system, which may consist of a number of partial digital twins that have been logically created to perform a variety of functions, as well as the device acquisition layer.

5.7 CONCLUSION

The proliferation of interactive Internet of Things devices has led to a considerable increase in the volume of multi-structured data that is produced, communicated, and kept in local and remote storage systems. The creation of digital twins for Internet of Things devices is currently in progress with the intention of better comprehending and articulating the structural features and behaviors of these devices in a range of scenarios. In fact, digital twins are able to simulate the behavior of physical objects in response to changing environmental conditions.

This chapter provided a comprehensive analysis of the myriads of potential ways in which the merger of these two technologies could bring a variety of digital advances as well as disruptions in the sector. This will make it possible for a variety of user cases, business cases, and technology cases to be envisioned and put into reality. In light of this, the field of digital twins is expected to witness tremendous growth in the years to come as a direct result of a cool synchronization with blockchain.

KEYWORDS

- **blockchain technology**
- **data transactions**
- **digital twin**
- **internet of things**
- **model integration**
- **simulation**
- **virtual twin**

REFERENCES

1. Minerva, R., Gyu, L., & Noel, C. (2020). Digital twin in the IoT context: A survey on technical features, scenarios, and architectural models. *Proceedings of the IEEE, 108*, 1785–1824.
2. Muneeswari, G., Sajithra, S., & Hegde, R. (2023). Self-diagnosis platform via IoT-based privacy preserving medical data. *Measurement: Sensors, 25*, 1–10.
3. Bazmohammadi, N. (2021). Microgrid digital twins: Concepts, applications, and future trends. *IEEE Access, 10*, 2284–2302.
4. Chen, G. (2021). Digital twin modeling for temperature field during friction stir welding. *Journal of Manufacturing Processes, 64*, 898–906.
5. Ogata, K. (2001). Introduction to digital twin: Modern control engineering. 4th ed.; Prentice Hall PTR: Upper Saddle River, NJ, USA.
6. Eckhart, M., & Ekelhart, A. (2018). A specification-based state replication approach for digital twins. *Proceedings of the 2018 Workshop on Cyber-Physical Systems Security and Privacy*, 1–7.
7. Schellenberger, C., & Zhang, P. (2017). Detection of covert attacks on cyber-physical systems by extending the system dynamics with an auxiliary system. *IEEE 56th Annual Conference on Decision and Control (CDC)–IEEE*, 1–6.
8. Vazhuthi, P., Aruchamy, P., & Manikandan, S. P. (2022). An energy-efficient auto clustering framework for enlarging quality of service in the Internet of Things-enabled wireless sensor networks using the fuzzy logic system. *Concurrency and Computation: Practice and Experience, 34*, 1–26.
9. Willard, J. (2022). Integrating scientific knowledge with machine learning for engineering and environmental systems. *ACM Computing Surveys, 55*, 1–37.
10. Abburu, S. (2020). Cognitwin–Hybrid and cognitive digital twins for the process industry. *IEEE International Conference on Engineering, Technology and Innovation (ICE/ITMC)*, 1–6.

11. Ritto, T. G., & Rochinha, F. A. (2021). Digital twin, physics-based model, and machine learning applied to damage detection in structures. *Mechanical Systems and Signal Processing, 155*, 107614.
12. Mohammad, A., Joonas, S., & Iivo, Y. (2021). Hybrid digital twin for process industry using Apros simulation environment. *26th IEEE International Conference on Emerging Technologies and Factory Automation (ETFA)–IEEE*, 1–6.
13. Mathupriya, S., Saira, B., & Sridhar, S. (2020). Digital twin technology on IoT, industries & other smart environments: A survey. *Materials Today: Proceedings*, 1–12.
14. Manso, G., & Gerrit, K. (2017). Physical model-based digital twins in manufacturing processes, Model-based control for smart forming processes. *Proceedings of 10th forming technology forum – University of Twente*, 1–7.
15. Ayani, M., Ganebäck, M., & Ng, A. H. C. (2018). Digital Twin: Applying emulation for machine reconditioning. *Procedia CIRP, 72*, 243–248.
16. Dai, S. (2021). Ontology-based information modeling method for digital twin creation of as-fabricated machining parts. *Robotics and Computer-Integrated Manufacturing, 72*, 102173.
17. Pylianidis, C. (2022). Simulation-assisted machine learning for operational digital twins. *Environmental Modelling & Software, 148*, 105274.
18. Leng, J. (2021). Digital twins-based smart manufacturing system design in Industry 4.0: A review. *Journal of Manufacturing Systems, 60*, 119–137.
19. Tao, F. (2018). Digital twin-driven prognostics and health management for complex equipment. *CIRP Annals, 67*, 169–172.
20. Angjeliu, G., Coronelli, D., & Cardani, G. (2020). Development of the simulation model for Digital Twin applications in historical masonry buildings: The integration between numerical and experimental reality. *Computers & Structures, 238*, 106–282.
21. Tao, F. (2019). Digital twins and cyber-physical systems toward smart manufacturing and industry 4.0: Correlation and comparison. *Engineering, 5*, 653–661.
22. Borth, M., Verriet, J., & Muller, G. (2019). Digital twin strategies for SoS 4 challenges and 4 architecture setups for digital twins of SoS. *14th Annual Conference System of Systems Engineering (SoSE)–IEEE*, 1–7.
23. Autiosalo, J., Siegel, J., & Tammi, K. (2021). Twin base: Open-source server software for the digital twin web. *IEEE Access, 9*, 140779–140798.
24. Cimino, C., Ferretti, G., & Leva, A. (2021). Harmonizing and integrating the digital twin's multiverse: A paradigm and a toolset proposal. *Computers in Industry, 132*, 103–501.
25. Romina, E., Francis, B., & Benoit, C. (2021). Conceptualizing digital twins. *IEEE Software, 39*, 39–46.
26. Zheng, P., & Abinav, S. (2020). A generic tri-model-based approach for product-level digital twin development in a smart manufacturing environment. *Robotics and Computer-Integrated Manufacturing, 64*, 101958.
27. Jiang, Z., Guo, Y., & Wang, Z. (2021). Digital twin to improve the virtual-real integration of industrial IoT. *Journal of Industrial Information Integration, 22*, 100196.
28. Semeraro, C. (2021). Digital twin paradigm: A systematic literature review. *Computers in Industry, 130*, 103469.

29. Sahlab, N. (2021). Knowledge graphs as enhancers of intelligent digital twins. *4th IEEE International Conference on Industrial Cyber-Physical Systems (ICPS)–IEEE*, 1–6.
30. Binsfeld, T., & Gerlach, B. (2022). Quantifying the benefits of digital supply chain twins—A simulation study in organic food supply chains. *Logistics, 6*, 1–16.
31. Chakraborty, S., & Adhikari, S. (2021). Machine learning-based digital twin for dynamical systems with multiple time scales. *Computers & Structures, 243*, 106410.
32. Chakraborty, S., Adhikari, S., & Ganguli, R. (2021). The role of surrogate models in the development of digital twins of dynamic systems. *Applied Mathematical Modelling, 90*, 662–681.
33. Mashaly, M. (2021). Connecting the twins: A review on digital twin technology & its networking requirements. *Procedia Computer Science*, 299–305.
34. Platenius-Mohr, M., File-and API-based interoperability of digital twins by model transformation: An IIoT case study using asset administration shell. *Future Generation Computer Systems*, 94–105.
35. Barbieri, G., Andres Gutierrez, D. (2021). A GEMMA-GRAFCET methodology to enable digital twins based on real-time coupling. *Procedia Computer Science*, 13–23.
36. Anbalagan, A., Shivakrishna, B., & Srikanth, K. S. (2021). A digital twin study for immediate design/redesign of impellers and blades: Part 1: CAD modeling and tool path simulation. *Materials Today: Proceedings*, 8209–8217.

CHAPTER 6

Blockchain-Based Digital Twins for Authentication and Remote Resource Sharing

S. JOTHI LAKSHMI,[1] R. PRADEEP,[2] K. VEENA,[1] and CHIEN THANG PHAM[3]

[1]*Department of Computer Science and Engineering, Akshaya College of Engineering and Technology, Coimbatore, Tamil Nadu, India*

[2]*Assistant Professor, Department of Electronics and Communication Engineering, Sri Ramakrishna Engineering, College, Coimbatore, Tamil Nadu, India*

[3]*TNU–University of Sciences, Thai Nguyen, Vietnam*

ABSTRACT

The digital twin (DT) of the diagnostic patients is created using augmented reality (AR) and maintained if a patient is facing an emergency during the unavailability of the doctor. The virtual item is sent by the already assigned doctor to the operating doctor for the operating doctor to grasp the internal status of the patient and treat the patient, among others. After a physical object is operational, a DT can be used for configuration, monitoring, diagnostics, and prognostics. The suggested approach also creates decentralized storage space to store, retrieve, and share DTs data. Here, DT is shared using the Ethereum Blockchain firewall Contract Account to secure data from third parties, and it is only accessible after the encryption process by the end user. As data in the medical field is hermetic, the

Blockchain-Based Digital Twins: Research Trends and Challenges.
A. Prasanth, Vandana Sharma, Rajesh Kumar Dhanaraj, Pethuru Raj (Eds.)

Double Advanced Encryption Standard (DAES) algorithm is introduced to create a strong ciphertext. By using the DAES algorithm, users can access the virtual object reliably after the authentication process. The highly privileged medical ledger is transferred in a regionalized manner, and it assists in saving many diagnostic patients.

6.1 INTRODUCTION

A wide range of industries, including business and industry, have noted significant developments in the area of digital twin technologies. With genuine data produced by real assets, it is feasible to recreate occurrences in a virtual scene [1]. This data is related to a remote server that uses digital twins to replicate events. Following that, the simulation outcomes may be made available to the data holder or even other additional users.

One of the key challenges is the need for accurate and up-to-date data. In many cases, the data used to create the digital twin is inaccurate or out-of-date. This can lead to inaccurate results from the simulation. Another difficulty of digital twins has the potential to transform how we do business. By replicating real-world events in a virtual scenario, we can gain valuable insights that would otherwise be unavailable. However, the implementation of digital twin technology is not without its challenges. In this chapter, we will explore some of the difficulties involved with using digital twins.

It is typical for organizational demands and people's expectations to vary and develop over time. They have been gracefully and easily met by breakthrough technology, both individually and jointly. As Internet of Things (IoT) devices become more popular in our daily lives, there is an increasing need to equip them with the requisite intelligence to enable flexible action and reaction [2–4]. Making connected gadgets intelligent is the goal. The greatest way to understand anything is to turn the data that has been obtained into knowledge and information. The changeover process is sped up by a few technological solutions, like integrated data analytics platforms, machine, and deep learning algorithms, etc. [5].

6.2 RELATED WORKS

Several methods and techniques have been strongly advocated by many for making smart devices. One of them is the development of digital twins for IoT devices. Digital twins continuously gather useful information from their corresponding physical twins and process it to extract actionable insights from data masses. The learned new knowledge is relayed to the physical tools to properly energize them for accomplishing their tasks. They learn to adapt their behavior wisely to the rapidly changing environment thanks to this ongoing nurturing [6].

One of the primary difficulties is deciding how to safely convey simulation data in real-time and how to disseminate simulation data. Sensitive information from data owners may be contained in data generated by physical assets, and if this data is leaked to an enemy, serious privacy issues may arise. Consider exchanging data with other data users to maximize the usability of data from digital twins. To overcome these problems, we suggest an architecture of the system for the appropriate exchange of information on digital twins [7]. The proposed system concept makes use of blockchain technology for data verification and cloud computing for effective data exchange.

An artificially created human being, typically in the form of a virtual, is called a digital twin. This image can be used as a patient stand-in in a medical setting or for research reasons. The development of a digital twin has the potential to improve patient care and research in the healthcare sector [8]. It can aid in decision-making regarding how to utilize equipment and what repairs are required by giving a better understanding of how it is functioning. For example, by producing a precise clone of a patient's brain, researchers can study illnesses more fully and understand how factors impact living tissue. By using digital twins instead of actual patients, hospitals can save money on labor costs and research programs. Healthcare professionals are currently talking a lot about digital twins [9]. They help hospitals improve patient care and research while also lowering the cost of research programs. Digital clones are identical duplicates of physical items or infrastructure, including a person's mind or an engine from a vehicle. Models are produced using computational modeling and aid researchers in understanding how systems operate by permitting viewers to witness exactly which specific factors were modified [10].

By utilizing a virtual model that accurately duplicates the individual's neurological makeup, investigators may examine different therapies on an individual's brain without administering any medications. Digital twins have recently begun to play the role of actual patients in the research of diseases such as cancer, which has long been the focus of academics

[11]. By conducting clinical trials with virtual patients rather than actual patients, doctors can save money on resources, labor costs, and research initiatives. To avoid investing money in new resources before knowing if they will be efficient enough to be used in actual patient care, hospitals may use digital twins to replace existing personnel during operations [12].

Digital twins enable digital patient management in life-threatening circumstances anywhere in the world. Although creating a virtual twin is a challenging task, safety while sharing the twin must be given top consideration. The blockchain firewall operates as a backend to conceal information from middlemen, and our virtual twins support honest protection for patients [13].

The following are a few advantages of creating in the medical field:

1. **Enhanced Precision:** Compared to using physical specimens, scientists may investigate disorders more precisely by generating a digital clone of a patient's brain.
2. **Reduced Labor Expenses:** Healthcare facilities might reduce expenditures on staffing and investigation activities by employing digital twins instead of real individuals.
3. **Better Communication:** Compared to employing human equivalents, medical professionals can communicate with one another more effectively and efficiently when using digital twins.

6.3 SUGGESTED APPROACH

The utilization of digital twin technologies in the real world is not always without complications. Finding a method to accurately convey and modify simulation facts in real-time is one of the main challenges. The information generated by physical property may also include sensitive information from information owners, and if this information is disclosed to an enemy, serious private difficulties may arise. The suggested machine concept makes use of blockchain technology for data verifiability and cloud computing for effective fact exchange.

6.3.1 SYSTEM MODEL

The system model introduces the network and threat models as demonstrated in Figure 6.1.

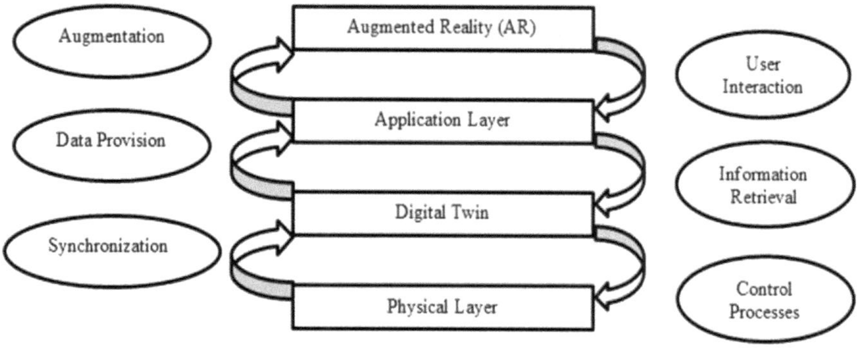

FIGURE 6.1 System model.

6.3.2 NETWORK PARADIGM

A blockchain network, or technical infrastructure, provides contract (chain code) services and an application ledger. The primary purpose of smart contracts is to establish transactions, which are subsequently transmitted to all peer nodes in the network and permanently recorded on each of their respective ledgers. Users of applications could be blockchain network administrators or end users utilizing client applications [14]. Typically, a collaboration of organizations forms the network, and when the network is initially configured, the collaboration decides on a set of regulations that will govern which organization's permissions are granted access to the network.

The described approach gives a network paradigm that utilizes blockchains for cloud-based digital twin scenarios as stated in Figure 6.2. The proposed concept consists of four parts: a blockchain, a cloud server, a client side, and a data consumer. A more thorough explanation of each entity can be found below:

1. **Cloud Server:** The web server has the processing power to run digital twin simulations and retain the outcomes. Following a mutual key agreement, Data owners' physical assets are accessed by the cloud server. The information is used by a cloud server to create virtual versions of digital twins. The findings of the simulations are subsequently communicated to the record via the server in the cloud. The server located in the cloud will also be able to supply data to customers' information on virtual twins according to the consent.

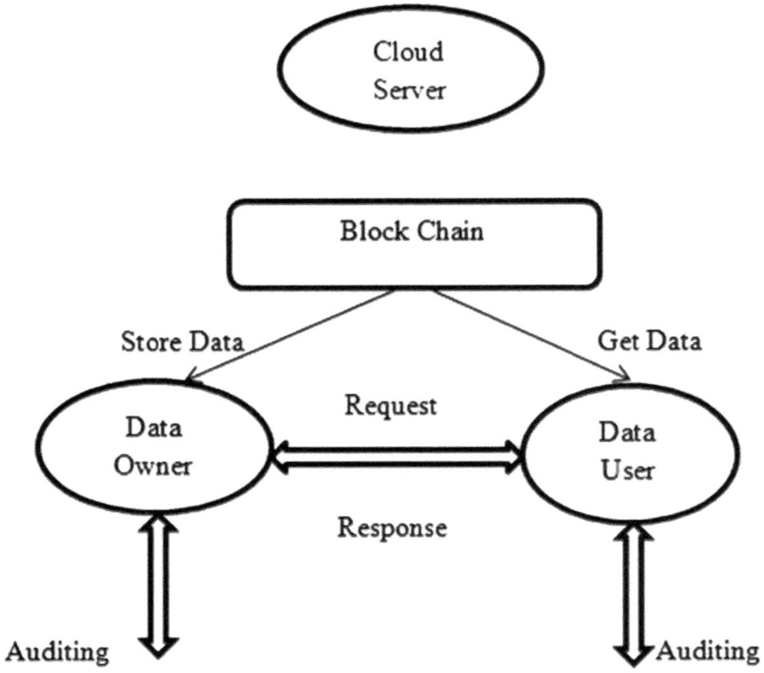

FIGURE 6.2 Network paradigm.

2. **Data Owner:** The owners own tangible objects like sensors and wristbands. The data owners use a smartphone or other device to gather the information provided by the assets, cross-check it with one another, and then upload the validated data to a cloud server. Furthermore, grants the cloud permission to provide the required data to the user upon request. The owner of the data can utilize the blockchain to inspect the log record of the data exchange after it has taken place.
3. **Data User:** These periodically seek access to the digital twin data. Data can be used by manufacturers, researchers, and technicians. An individual user can access the stored digital data after successfully verifying their identities. The individual can then obtain the required data and use the blockchain to verify its accuracy.
4. **Blockchain:** This maintains hash values for data saved on the cloud database as well as log entries for the consumers [15]. Owners

utilize log records to verify information has been distributed to the authorized customers. Individuals, on the other hand, utilize the hash function to ensure that there is no change in the information. The blockchain saves hash values for data saved on the storage in addition to logs between the cloud and consumers. Blockchain is the technology that powers cryptocurrencies like Bitcoin and Ethereum and it is quickly becoming a popular mechanism for the secure storage of arbitrary data [21–23].

6.3.3 THREAT MODEL

Blockchain technology is a cutting-edge innovation. Exploring the security of this innovative technology is essential due to its potential. Although a lot of work has been done in this area, no definitive threat model that labels all potential dangers and attack vectors within the blockchain ecosystem is now available. It is quite helpful to have a framework to utilize in identifying known attacks and pointing out ones that may have been overlooked when talking about potential security threats to a system and trying to determine whether a system is secure by design. To establish a useful threat model for the blockchain, the security risks associated with it are mapped to STRIDE, a well-known threat model created by Microsoft. Microsoft created the STRIDE framework to aid with threat modeling. Each letter of the acronym STRIDE stands for one of the most prevalent cybersecurity threats:

1. **Spoofing:** The capacity of the attacker to pose as someone else on the system is referred to as spoofing.
2. **Data Integrity:** It is violated by tampering attacks on the protected system's data.
3. **Repudiation:** The capacity to retract one's previous statements or actions is known as repudiation.
4. **Information Disclosure:** This category includes violations of confidentiality.
5. **Elevated Privileges:** A privilege escalation attack occurs when a user can take over the system with unlawful levels of authority.

In the context of blockchain technology, we can categorize elevated privileges according to whether the attacker has unauthorized access to a user's account, a significant amount of control over the blockchain system (as in a 51% attack), or unauthorized permission access to a smart contract. For describing the probable effects that specific flaws or attacks may have on the security of a system, the STRIDE framework is helpful. Blockchain systems, on the other hand, are a full environment that includes everything to increase the functionality of the system and support its security. The suggested method analyzes the safety using the Dolev-Yao (DY) threat prototype model. According to this paradigm, opponent A is capable of a number of things.

Opponent A has total authority over wireless channels and can track or eavesdrop on them. Using a power analysis attack, opponent A can acquire the values stored from a card reader owned by an authorized user. Because opponent A has access to a dictionary of IDs and passwords, it may attempt to guess both a genuine user's password and identity at the same time. An intruder may attempt to exploit security flaws such as session key disclosure and MITM.

6.3.4 SYSTEM MODEL PHASES

To provide safe authentication processes for systems that exchange blockchain- and cloud-based digital twin data. The stages of the proposed approach are verification of information, owner cloud verification, ownership-user identification, the initialization process, authorization, and ownership. Users, cloud servers, and data owners are registered on the network that B establishes.

Data owners then create a secure connection with cloud servers and send real-time data produced by their physical assets. Using the digital twins of the assets, cloud servers may replicate the assets. The generated simulation results are stored in the blockchain together with the resulting data hash values. When a data user requests information on a digital twin, the user must first establish an authentication relationship with the data owner. The log records are kept after data has been traded and then saved in the blockchain. Below, each stage is described in more depth.

6.3.4.1 INITIALIZATION

At the beginning of the process, B publishes the network's public keys. To create S.Z, B selects a large prime pK and also selects the two multiplicative groups D and DT of order pK. B selects e: GXG → DT, X(.): {0,1}* → D, x(.): {0,1}* → Z, a generator P_K of D, and computes PTA = SCA. P_i. B publishes D, DT, P_i, PTA, pK, X(.), and x(.). while keeping S private (Table 6.1).

TABLE 6.1 Notation and Description

Notation	Description
DOwj	j^{th} owner of the data
DOUk	k^{th} owner of the data
CSj	j^{th} owner of the data
ID	Identity of DOwj
PW	Password of DOwj
HID	Pseudo identity of DOwj
SID	DOwj's secret identity
Sa, Sb	DOUk and DOwj – secret key
U, Uz	Generation of random numbers in session
Rq	DOUk 's request message
La, Lb	A message digest of DOUk
Ra, Rb	The shared Diffie-Hellman key between DOwj and CSj
Y1, Y2, Y3, Y4	Timestamps
||	Concatenation operation

6.3.4.2 REGISTRATION

To access the network, each entity must first register with the B. Before DOw_j registration, B constructs sj and ID and gives them to Pj sj and CS_j. The public and private keys of CS_j are denoted by P and sj, respectively. DOwj can register with B utilizing its specific smart gadget SDi, which can send the created data to CSj. DOw_j then transmits ID, HID to B after calculating HID = X (ID||PW||a_i). B first checks to see if an ID is already registered. If it is not registered, the fuzzy verifier l 25, 210 generates S_a, and before passing SID, si, l to DOw_j, it calculates SID = S_a.HID and Pi = si. P_i. Doi calculates:

$$X_i = x_i \text{ HPW};$$

$$Y_i = S_a \text{ h}(x_i||\text{HPW});$$

$$Z_i = \text{SID h}(S_a \,||x_i|\text{HPW}); \text{ and}$$

$$A_u = \text{h}(S_a \,||x_i||\text{SID}).$$

6.3.4.3 CLOUD OWNER AUTHENTICATION

The ability of CS_j to communicate data generated by their physical assets should be confirmed by DOw_j. In response, SDi computes HPW = x(ID||PW), $x_i = X_i$HPWi, $S_a = Y_i$ h(x_i||HPW), and SID = Z_ih(S_a ||xi||HPW), and verifies A_u = h(S_a ||x_i ||SID). DOw_j then passes ID and PW to SDi. If they are equal, SDi determines HID = X (ID||PW||x_i), Ri = r_i.x_i.P, $R_a = r_i$. x_i.Pj, PID = HID h(R_a|| Y_1), and X_i = SID.h(HID||R_a|| Y_1). Then, DOw_j transmits to CSj R_i, PIDi, X_i, and Y_1. Once the message has been received, CSj calculates R_a = sj. R_i and HID have the same value as PID h(R_a|| Y_1) and e(HID.h(HID||R_a|| Y_1.PTA)) e.(X_i, P).

If they are the same, CS_j produces r_j, and Y_2 calculates Rx = r_j. $P_i = r_j$ and R_b. R_i, S_b, and L_b are equal to h(R_a||R_b||HID). Then, CSj sends Y_2 to R_x L_b. After receiving the communication, DOw_j determines R_b=r_i.a_i. L_b equals h(S_b||R_a||R_b), R_x equals h(R_a||R_b||HID), and S_b equals h(R_x||R_b||R_a). The data produced by DOw_j 's physical assets are then encrypted and sent to CSj, which simulates using the data there before sending the results back to DOw_j. The data compiled using the hash function in CS_j is then uploaded to the blockchain together with the authentication signature of the DOw_j corresponding to the information [19].

6.3.4.4 AUTHENTICATION OF USER-OWNER

When necessary, DOU_k can ask DOw_j for information from its digital counterpart. The request message Rq=Z is created by DOU_k using the random number U_z and the date Y_3 as inputs. Then DOU_k calculates U_z.ak. P_i=U_z and U_z.a_k. Pi equals U_z, PID_k, HID_k, and h (U_{ba} Y_3), M_k, X_k, and

SIDk.h (HIDk, Rqk, and U_{ba}, Y_1), and sends U_z, PID_k, M_k, X_k, and Y_3. In this stage, DOw_j authenticates DOU_k and communicates with DOU_k in response. DOU_k sends a message to DOw_j, who then calculates $U_{ba} = S_b$. e(HIDk.h(HIDk ||Rq $||U_{ba}||$ Y_3), PTA) = e (X_k, P), U_z HIDk = PIDk h($U_{ba}||$ Y_3), Rq = M_k h(HIDk $||U_{ba}||$ Y_3), and. If they are the same, DOw_j generates ui at arbitrary and determines U=ui. P, $S_b$ik=h($U_{ba}||U_{ab}||$HIDk||HID), and L_a=h(SDik$||U_{ba}||U_{ab}$). U, L_a, and Y_4 are subsequently sent from DOw_j to DOU_k, who calculates U_{ab}= U_z.ak. HI, U, $S_b$ik (U_{ba} U_{ab} HIDk HID) and L_a=h tests ($S_b$ik U_{ba} U_{ab}). At last, DOw_j and DOU_k settle on using the same private key, $S_b$ik. CSj uses the key to encrypt DOw_j 's data and deliver it to DOU_k after receiving $S_b$ik from DOw_j. To check the accuracy of the data, DOU_k can decode the hash values from the blockchain.

6.4 DISCUSSIONS

6.4.1 PERFORMANCE ANALYSIS

The proposed method evaluated the proposed authentication strategy against newly reported techniques [16, 18, 20] in related scenarios.

1. **Computational Cost:** It is viewed as a key breakthrough by the blockchain community. The "proof of work" and "proof of stake" consensus algorithms are greatly improved as a result, which creates a transparent and efficient framework for mining operations. Smart contracts use the computational cost, which is an Ethereum unit, to pay miners as part of the consensus procedure because miners are essential to adding and validating transactions to the blockchain. Consequently, the pricing of computational expenses may differ depending on the processing costs necessary to complete the transaction and the amount paid.

   The entire cost is also influenced by the smart contract's complexity, frequency, and intensity. The contract will fail if miners are not paid enough ether to complete the transaction, and any gas consumed will be returned. The Ethereum platform was developed by Vitalik Buterin, who based the use of computational charges on the economics of scarcity as a mechanism to reduce the computational waste of transactions using extra computer resources. The suggested approach is based on research done in

Ref. [16], which made use of the Java pairing-based cryptography package. On a computer with a quad-core Intel i7-8300H CPU operating at 2.3 GHz and 16 GB of RAM, the experiment was conducted, and the following amounts of time were required for each operation (Table 6.2):

- **Xbp:** The bi-linear pairing operation's computational cost ≈ 17.4 ms.
- **Xhp:** The hash procedure was map-to-point with the cost of computation ≈ 42.10 ms.
- **Xexp:** A modular exponentiation operation's computational cost ≈15.80 ms.
- **Xmul:** A computing cost of performing point scalar multiplication≈ 13.50 ms.
- **Xadd:** A point addition operation's computational cost ≈ 0.48 ms.

TABLE 6.2 Comparison of Computation Cost

Scheme	Computation Cost
Wu et al. [16]	387.46 ms
Khatoon et al. [18]	347.1 ms
Sengupta et al. [20]	227.48 ms
Proposed	193.76 ms

Due to their low computational demands, the suggested approach ignores the expenditure of the EX-OR function and a one-way hashing value. It examined the duration of the sign-in and authentication procedures. At first, the Wu et al. [16] scheme's total cost of computation is 2Tbp 11Tmul+4Thp+2Tadd387.46ms. Second, the Khatoon et al. [18] scheme's total computation time is 4Thp 2Tbp+7Tmul 347.1 ms. The scheme's overall computation time is 4Tbp 4Tmul 2Thp+41T = 227.48 ms [20]. In addition, the proposed system has a time cost of 2Tbp+Thp+8Tmul 193.76ms. The above demonstrates how these costs rise with an increase in the number of authentications whereas the table compares computational costs. The suggested procedure has a lower computing cost than previously stated methods [16, 18, 20] as demonstrated in Table 6.2.

2. **Communication Cost:** This of the new approach and existing protocols [16, 18, 20] were examined as illustrated in Figure 6.3 and Table 6.3. The bit widths for the identification, a random integer, an expiration date, or a collection of characters in the result of an encryption hash, and G1 were found. Cost of computation as the number of authentications rises.

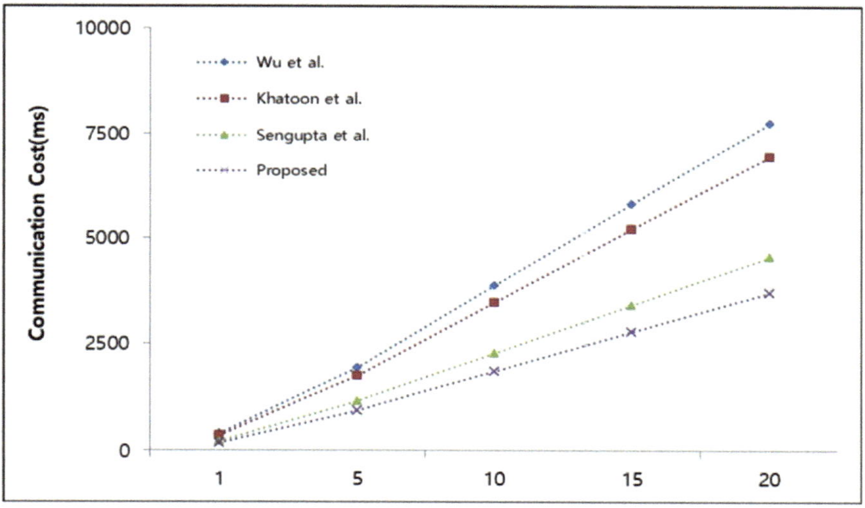

FIGURE 6.3 Computational cost.

In Ref. [16], the first message is (Idi, Rin, Ri1, Rsi, Y_1), while the next is (Id_j, Rj_n, R_x1, h_j), which contains a G1 set of five elements, an output of hash, two or more identifiers, and a timestamp. Communication costs a total 5120 + 256 + 256 + 32 = 5,664 bits. The first message of the Khatoon et al. scheme [18] is (R_i, T_i, A_u), and the next one is (R_s, T_s, A_u). The total bit cost of these messages is 2048 + 512 + 64 = 2,624 bits and contains two G1 sets of elements, two hash function outputs, and two timestamps.

TABLE 6.3 Comparison of Communication Cost

Scheme	Total Communication Bits
Wu et al. [16]	5,664 bits
Khatoon et al. [18]	2,624 bits
Sengupta et al. [20]	5,184 bits
Proposed	4,672 bits

Sengupta et al. [20] have two pieces of information, the foremost thing is (CID, N_i, Zi, Fi, Ti) and the next one is (Tss). It contains a total of 5120 + 64 = 5,184 communication bits because they have five G1 group components and two timestamps. The suggested scheme's first one is (U_z SID_k, M_k, X_k, Y_3) and the second one is (U, L_a, Y_4). A group of four elements, two timestamps, and two hash function outcomes sums to obtain the overall transmission cost rate of 4096 + 512 + 64 = 4,672 bits. The suggested approach will be much more effective than existing approaches [16, 20]. Further, the suggested protocol is more secure than the equivalent methods while having a little higher communication cost than the method suggested in Ref. [18].

6.5 CONCLUSION

In this chapter, a system model and a privacy-preserving communication mechanism were designed for a blockchain-based cloud-based digital twin environment. According to the suggested method, data produced by physical assets is gathered by a data owner's smart device and transferred to a cloud center point after mutual verification. The simulation information can then be securely communicated to the data owner, who will receive it in real-time. Before sending the desired information to the user via a server located in the cloud in response to an information request from a data consumer, the data owner validates the individual's identity. The blockchain allows the data user to access the information and confirm its accuracy. Through informal testing, we demonstrated the suggested technique's resistance to various attacks.

KEYWORDS

- **augmented reality**
- **authentication standards**
- **blockchain**
- **communication mechanism**
- **digital twin**
- **double advanced encryption standard**
- **encryption algorithm**

REFERENCES

1. Jia, P., Wang, X., & Shen, X. (2023). Accurate and efficient digital twin construction using concurrent end-to-end synchronization and multi-attribute data resampling. *IEEE Internet of Things Journal, 10*, 4857–4870.
2. Aruchamy, P., Sabeena, G., & Sowndarya, K. (2023). An artificial intelligence approach for energy-aware intrusion detection and secure routing in Internet of Things-enabled wireless sensor networks. *Concurrency and Computation: Practice and Experience*, 1–28.
3. Jayachitra, S., Prasanth, A., & Hariprasath, S. (2023). AI-enabled Internet of Medical Things in smart healthcare. In *AI Models for Blockchain-Based Intelligent Networks in IoT Systems: Concepts, Methodologies, Tools, and Applications* (pp. 141–161).
4. Dhiviya, S., Malathy, S., & Kumar, D. R. (2018). Internet of Things (IoT) elements, trends, and applications. *Journal of Computational and Theoretical Nanoscience, 15*, 1639–1643.
5. Pushpalatha, N., Sudha, M., & Sivaranjani, S. (2023). Enterprise data, data fabric architectures: Web-driven applications. *5*, 95–116.
6. Mengnan, L., Fang, S., Huiyue, D., & Cunzhi, X. (2020). Review of digital twin about concepts, technologies, and industrial applications. *Journal of Manufacturing Systems, 58*, 1–16.
7. Liu, B., Yu, X. L., Chen, S., Xu, X., & Zhu, L. (2017). Blockchain-based data integrity service framework for IoT data. *IEEE International Conference on Web Services, 1*, 468–475.
8. Wei, P. C., Dahu, W., Zhao, Y., Tyagi, S. K., & Neeraj, K. (2019). Blockchain data-based cloud data integrity protection mechanism. *Future Generation Computer Systems, 102*, 1–10.
9. Haiyan, W., & Jiawei, Z. (2019). Blockchain-based data integrity verification for large-scale IoT data. *IEEE Access, 10*, 1–11.
10. Alam, K. M., & Abdulmotaleb, E. S. (2017). A digital twin architecture reference model for the cloud-based cyber-physical systems. *IEEE Access, 14*, 1–18.
11. Liu, Y., Lin, Z., Yuan, Y., & Zhou, L. (2017). A novel cloud-based framework for the elderly healthcare services using a digital twin. *IEEE Access, 7*, 49088–49101.
12. Shohin, A., Xun, X., Ray, Z., & Yuqian, L. U. (2021). Digital twin as a service (DTaaS) in Industry 4.0: An architecture reference model. *Advanced Engineering Informatics, 47*, 1–15.
13. Wang, C., Cai, Z., & Li, Y. (2022). Sustainable blockchain-based digital twin management architecture for IoT devices. *IEEE Internet of Things Journal, 10*, 6535–6548.
14. Huang, S., Wang, G., Yan, Y., & Fang, X. (2020). Blockchain-based data management for digital twin of product. *Journal of Manufacturing Systems, 54*, 361–371.
15. Shen, W., Hu, T., Zhang, C., & Ma, S. (2021). Secure sharing of big digital twin data for smart manufacturing based on blockchain. *Journal of Manufacturing Systems, 61*, 338–350.
16. Wu, T. Y., Lee, Y. Q., Chen, C. M., Tian, Y., & Al-Nabhan, N. (2021). An enhanced pairing-based authentication scheme for smart grid communications. *Journal of Ambient Intelligence and Humanized Computing, 5*, 1–13.

17. Chen, Y., Martinez, J. F., Castillejo, P., & López, L. (2019). A bilinear map pairing-based authentication scheme for smart grid communications. *IEEE Access, 7*, 1–11.
18. Khatoon, S., Rahman, S. K., Alrubaian, M., & Alamri, A. (2019). Privacy-preserved, provable secure, mutually authenticated key agreement protocol for healthcare in a smart city environment. *IEEE Access, 7*, 47962–47971.
19. Nikooghadam, M., & Amintoosi, H. (2019). A novel provably secure ECC-based authentication and key management protocol for telecare medical information systems. *IEEE Access, 5*, 1–18.
20. Sengupta, A., Singh, A., Kumar, P. J., & Dhar, T. (2022). A secure and improved two-factor authentication scheme using elliptic curve and bilinear pairing for cyber-physical systems. *Multimedia Tools and Applications, 81*, 22425–22448.
21. Son, S., Lee, J., Park, Y., Park, Y., & Das, A. (2022). Design of blockchain-based lightweight V2I handover authentication protocol for VANET. *IEEE Transactions on Network Science and Engineering, 9*, 1–18.
22. Thang, P., Trang, T., & Tai, B. (2022). Multimedia privacy, security, and protection within the blockchain: A review. In *Proceedings of the 5th IEEE International Conference on Contemporary Computing and Informatics* (pp. 1171–1176).
23. Hasan, H. R., Salah, K., Jayaraman, R., & Omar, M. (2020). A blockchain-based approach for the creation of digital twins. *IEEE Access, 8*, 34113–34126.

CHAPTER 7

Comprehensive Analysis of Industrial Internet of Things (IIoT)-Enabling for Digital Twin and Blockchain

MANI DEEPAK CHOUDHRY,[1] S. JEEVANANDHAM,[1] M. SUNDARRAJAN,[2] J. DHARANI,[3] and AKSHYA JOTHI[4]

[1]*Department of Computing Technologies, SRM Institute of Science and Technology, Kattakulathur, Tamil Nadu, India*

[2]*Department of Computer Science and Engineering, SRM Institute of Science and Technology, Ramapuram Campus, Chennai, Tamil Nadu, India*

[3]*Department of Information Technology, Dr. N.G.P. Institute of Technology, Coimbatore, Tamil Nadu, India*

[4]*Department of Computational Intelligence, SRM Institute of Science and Technology, Kattankalathur, Chennai, Tamil Nadu, India*

ABSTRACT

The term "industrial internet of things" (IIoT) is used to describe the interconnection of various sensors, instruments, and other equipment over computer networks used in many industrial contexts, such as but not limited to production and energy management. The ability to gather, share, and analyze data is a key enabler of productivity and competency gains, among other economic advantages. Cyber-physical systems, the cloud, big data analytics, Artificial Intelligence (AI), and Machine Learning (ML)

Blockchain-Based Digital Twins: Research Trends and Challenges.
A. Prasanth, Vandana Sharma, Rajesh Kumar Dhanaraj, Pethuru Raj (Eds.)

are just some of the technologies that make the IIoT possible. With the help of the IoT, AI, ML, SA, and spatial network graphs, digital twins may create dynamic simulation models that evolve and adapt in real-time to the state of their physical counterparts. A digital twin is a copy of anything that is continually learning and updating its characteristics based on data from a diversity of causes to reflect the near real-time state of the original thing. This system is capable of learning from a variety of sources, including sensor data expressing numerous facets of its functioning state, anthropological professionals like technologists with extensive and pertinent business field acquaintance, alike machineries, related task forces of machineries, and grander arrangements and situations. A digital twin is a programmed depiction by means of a machine that also includes the use of data from the machine's prior lives. For manufacturers, digital twins mean cutting expenses, keeping tabs on assets, improving maintenance, cutting down on downtime, and opening the door to the development of linked goods. This chapter analyzes the various technologies, architecture, and industrial revolutions that makeup IIoT.

7.1 INTRODUCTION

The era of revolution in industry first began in the 1780s with the expansion of systems to generate electricity from thermal energy, hydropower, and biofuels. During the 1870s' second industrial revolution, electrical energy was favored by factories and other industrial establishments. During the era of the 3rd uprising of industry, which began in the 1970s, embedded hardware, and computing technologies were used to introduce the notion of automation to the manufacturing industries. Industry 3.0, also known as the computerized age, was widely adopted as a result of technological advancements starting from 1970. However, with the rise of the IoT, Industry 4.0, also called the connected age, has reached its full potential. The Fourth Industrial Revolution was a watershed moment for the IoT. These years of the industrial revolution are sometimes referred to as the "connected age" [1].

The advent of the IoT and soft computing laid the groundwork for semantic interoperability, which is the foundation of Industry 4.0 since it allows for a seamless flow of information between the digital and physical spheres in real time [2]. There are many technological pioneers and business leaders who are excited about the arrival of the Fifth Industrial Revolution (IR 5.0), despite the fact that Industry 4.0 is just getting started [3–5]. Due to increased connectivity and cognitive technology, Industry 4.0 forecasts a rapid technological transformation in business and social practices [6]. Expert systems [7], cyber-physical systems [8], IIoT [9], cloud-based services [10], indoor industrial units, IoT, and AI [11] are examples of the latest technological developments that contribute to designing Industry 4.0. The era of Industry Uprising is depicted in Figure 7.1.

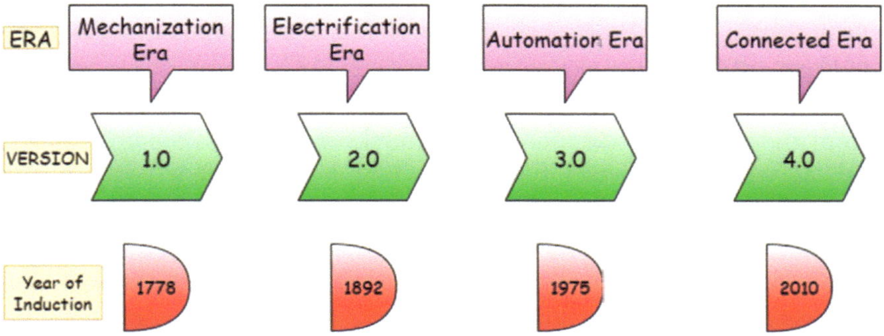

FIGURE 7.1 Revolution of industry automation.

Industries have been striving in recent decades to keep up with the market's shifting needs. Production facilities must be flexible, intelligent, and dynamic in order to meet client demands. Executives from corporations and production companies believe that commercial and industrial activities should be integrated. Such a merger suggests considerable advancements in business tactics and procedures. Additionally, this can only be done by combining various organizational components like suppliers, production facilities, and clients. One of the biggest developments of Industry 4.0, according to some researchers, was the origin of the IoT.

In comparison to worldwide ecological stewardship, Industry 4.0 innovations have given more emphasis to improving the relevant machine learning algorithms in order to boost efficiency and precision. On the other hand, IR 5.0 links ecological and social control with computing technology, making it simpler to put the necessary adjustments into place to tackle climate change. Distributed invention is a form of devolved manufacturing used by businesses using automation; it is often discussed in the context of industrial automation and advanced manufacturing [12].

Constructing a communication mechanism for sharing information, evaluating performance, and making decisions in real-time as a group is what this requires.

7.1.1 BLOCKCHAIN TECHNOLOGY

When neither verification nor audit methods are present, the issue of confidence in information systems becomes incredibly complex, especially when they deal with sensitive data like financial transactions involving virtual currencies. Blockchain is the system that enables a collection of untrustworthy parties to verify transactions [13]. It provides a decentralized, public, auditable, and unalterable ledger. All transactions made from the system's inception are available on the blockchain, and they can be verified and assembled at any time by any authorized party. The blockchain protocol organizes data into a series of blocks, each of which contains a collection of Bitcoin transactions that took place at a specific moment. A reference to the preceding block forms a link between blocks, making a chain.

A smart contract is a set of procedures of a system or code that allows a contract to be inevitably completed or obligated based on a set of agreed-upon conditions. In the case of Bitcoin transactions, for instance, intelligent control defines the underlying application logic that must be implemented. Since smart contracts enable the specification of criteria and functions that go beyond the switchover of cryptocurrency as the endorsement of assets in a particular choice of non-monetary associates, they are the appropriate component for expanding the usage of blockchain technology into new industries.

7.1.2 DIGITAL TWIN

At the forefront of the smart manufacturing transition is Digital Twin, which is enabled by the IoT and robust data rationale. Thanks to IoT, the amount of data available for usage in sectors like manufacturing, healthcare, and smart city planning has increased dramatically. Support for irregularity findings in medical services, malfunction exposure, and traffic administration in a smart city [14] are just a few examples of how the IoT's specific background paired with data rationale can be a valuable resource for a wide range of applications.

We are trying to find solutions to the following study concerns by performing this review:

- **Q1:** What is a "digital twin," and what are some common misunderstandings about it based on prior and current definitions?
- **Q2:** What are the uses and difficulties of enabling technologies?
- **Q3:** Does there exist any relation between Digital Twins and IoT, IIoT?
- **Q4:** What are the problems and exposed queries around digital twins?

The rest of this chapter is systematized in the following manner. The most current research on blockchain-based technologies is covered in Section 7.2. The main enabling technologies are outlined in Section 7.3. A comparative analysis of the most current studies is summarized in Section 7.4. Finally, Section 7.5 serves as the chapter's conclusion.

7.2 BACKGROUND

As far as we are concerned, a thorough analysis of the design and application issues of DTs based on digital ledgers has never been performed. However several subsequent studies have emphasized the application of blockchain in DTs. The authors of the most comprehensive study [15] have developed taxonomies of DTs, shedding light on their development processes, design iterations, enabling methodologies, and potential for fundamental needs. The author analyzes the advantages of adopting blockchain for DTs and provides case studies of DTs and collaboration in the automotive sector.

Concurrency, data privacy difficulties, resource-constrained device issues, and compatibility issues brought on by multiple consensus processes, negotiation systems, and smart agreements are all discussed as potential roadblocks to the widespread use of blockchain technology in DTs. In order to protect DT's organizational and operational data, it also discusses security components like privacy and integrity, but it ignores other safety-related design details like security precautions and part immunity. The publications reviewed in the categorical review are referenced in the following sections, which go into greater depth about each paper and emphasize any themes and case studies that were used. The divisions, which are not restricted to but include medical services, intelligent environment, and manufacturing, will discuss the key issues of the papers collected in those areas. The presentation order reflects the current status of the field and is based on the number of articles discovered in each category.

7.2.1 HEALTHCARE

The uses and advantages of DTs are examined in a number of industries, including manufacturing, autos, logistics, services, aviation, and healthcare, in an updated conceptual essay [16]. The project's objective is to create blockchain-based trust tokens (DTs), which facilitate data tracing and numerical identification through blockchain tracking. The authors of Ref. [17] examine how blockchain and distributed ledger technologies work together and are impervious to error for the performance of beneficial life processes. They also offered an illustration of the blockchain-based DT system developed utilizing literature reviews, market research, and plotting of high-tech real-world scenarios for DTs, IoT, blockchains, and intelligent contracts. Additionally, this study examines possible solutions to the combination problems that blockchain-based DTs encounter.

Focus on distinct blockchains, for instance, to facilitate communication and coordination among various system components during network scalability and the diagnosis and remediation of errors. However, the study does not discuss work on DTs based on blockchains explicitly. Similar to this, a blockchain-focused study [18] focused less on DTs' perspectives on using blockchain in manufacturing, while a DTs-focused poll [19] only mentioned blockchain mechanization as a way to address issues with information trust.

The chapter [20] explains how HP and AI/IoT specialists worked together to create human "Digital Twin" avatars. By examining data from both sources, strategies of Digital Twin and techniques of AI may be used to forecast the results of making specific changes to a person's lifestyle. This application illustrates the full fusion of information from the human and digital twins (The Replica). Allowing people to visualize how their choices affect their physical twin and shedding light on how their new habits could affect them.

By contrast, the chapter [21] shows how to create a patient digital twin with the use of Industry 4.0 and the Internet of Things to perform a practical prototype of autonomous surgery. The authors propose carrying out surgical procedures remotely through a mobile network. An advanced robotic arm, VR, and a 4G network allow the prototype to carry out accurate surgical procedures. The complexity of cross-disciplinary studies is acknowledged in the paper as a justification for the use of simulation in place of a physical prototype. Writers also discuss issues with syncing a model with a digital duplicate. Researchers take a look at how progressions in the domains of enabling technologies have helped comfort the problems of connection, incorporation, and cross-disciplinary study.

7.2.2 SMART CITIES

Urbanization has significantly increased in recent years, along with the growth of IoT and data analytics, according to research [22]. This is one of the driving forces for the work of Ref. [23], who also acknowledge it as inspiration. They describe the various states of spatiotemporal flux and stress the importance of understanding them in order to sustain progress. The idea they describe does this through the use of virtual reality headsets and Digital Twins, enabling them to track fluctuations and make forecasts using real-time analytics.

The work stated in Ref. [24] offers a framework, my SMART life, which can be used to create a digital duplicate of a smart city by using IoT advancements in several cities. This chapter offers a scenario on the use of the Digital Twin in supporting urban development and built environments, with a focus on its applications in the energy consumption sector, due to its ability to monitor and compare energy use based on environmental and human influence. Both are in use for present and future enhancements.

The chapter [25] examines and creates a case study that looks at how wind farms are used and how much energy they utilize for a smart grid. The report lists a few difficulties with wind farm servicing in particular and the dependability of power consumption. The researchers offer a way of operation that employs IoT sensors along with business intelligence in a Digital Twin setting to precisely carry out and forecast operations on the wind generators.

7.2.3 MANUFACTURING

There aren't many evaluations in this field, but Qi & Tao [26] provides a complete "360" viewpoint for big data associated with digital twins in engineering and industrial scenery, making it stand out as the first one worth noting. It analyzes the many technologies that enable Digital Twins and argues that these cutting-edge tools are crucial to the development of smart manufacturing. Production may be optimized with the use of machine learning algorithms and "virtual clones" of a system. This study classifies important aspects of industrial development in order to come up with their aims and ideas for applying a digital twin to the manufacturing system. In this way, companies may evaluate, repeat, and perfect their manufacturing procedures in a virtual environment, resulting in greater efficacy and efficiency [27].

According to Xu et al. [28], there are issues with the availability of samples for generating exact AI algorithms for new industrial processes, which is something that is also necessary for developing precise Digital Twins. He recommends using AI to help with problem diagnosis in a smart manufacturing setting. To guarantee that problems are kept to a minimum, the model's front-running makes use of a Digital Twin, which can learn and diagnose mistakes while also supplying training data. The first step of the method collects training data, which may subsequently be used for transfer learning in the second phase. A Digital Twin and transfer learning is used to enhance a fault-diagnosis system. The notion is put to the test and analyzed by the writers in a real-world case study that has to do with the production of autos.

Another essential area for implementing the usage of digital twins is the implementation of manufacturing procedures and systems throughout the design and development phase. Chhetri et al. [29] suggests a method that utilizes Digital Twins and the current interest in Cyber-Physical Systems (CPS). Researchers provide a Hierarchical Digital Twin Framework for the development and upkeep of adaptable CPSs inside the framework of smart manufacturing. All along the CPS design process, the Digital Twin makes use of its capability to recreate and check against real data on a virtual twin. This setting is ideal for seeing the digital twin merits since it allows the authors to use the coincident twin anticipatory conceptual representation and conduct training on a large scale.

The intelligent production design is credited as being a result of technological advancements in automation, as stated in Ref. [30]. The authors offer a methodology and conduct an analysis to leverage the growth of the Internet of Things to develop a digital twin of a manufacturing system. This approach takes advantage of IoT sensors to emphasize an operation's noise, power, and magnetostrictive output in a roundabout way. These aid in finding problems in manufacturing and identifying errors. The collective evaluation was based on a demonstration of a Fused-Deposition Modeling system that generates high-accuracy detection systems. In order to achieve national and international targets, it is necessary to assess and reduce industrial energy consumption, which in turn necessitates the development of more energy-efficient production systems.

In Ref. [31], we see the concepts and infrastructure for low-energy manufacturing. The positive effects on the planet's ecology are a driving force behind this. Costs will go down, earnings will go up, and new investments will be made as a result of more efficient manufacturing. The author constructs an energy-conscious Digital Twin model using an architecture that makes use of the MCloud platform. Together, they contribute to the development of what is known as Industry 4.0, a setting in which CPSs and Digital Twins are used to continuously monitor and self-configure production processes with the aim of optimizing energy efficiency.

7.3 ENABLING TECHNOLOGIES, ITS METHODOLOGIES, RESEARCH OPPORTUNITIES, AND CHALLENGES

The rise of IoT technology is evident in the number of devices that have been registered annually. The amount exceeded 17 billion in 2018. Over 75 billion gadgets are expected to be in use by 2025, and the market is expected to be valued at over $5 trillion. The notion of a fully networked world is supported by a sizable number of connected gadgets. This idea of connected services using IoT is depicted in Figure 7.2. IoT device development benefits everyone and has an influence on a variety of ordinary routines, including the communication industry, medicine, architecture, transportation, smart cities, and production [32].

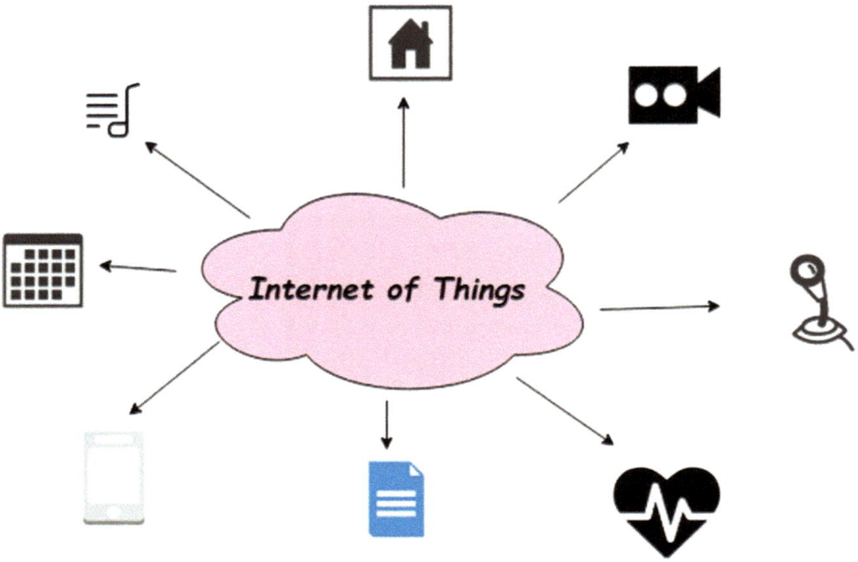

FIGURE 7.2 IoT devices.

7.3.1 IIOT

Many researchers have different interpretations of both the IoT and IIoT concepts [33]. Having many of the same characteristics as the IoT, but with an emphasis on manufacturing instead. Initial systems utilized in manufacturing and industrial settings are known as Industrial Control Systems. As things become more intelligent and autonomous, the IIoT may help us reap the benefits. Unlike IoT and IIoT, CPS are distinct but related technologies [34]. Just as IoT has the potential to significantly impact everyday life, IIoT has the potential to have a profound effect on manufacturing processes by facilitating the evaluation of tasks with more information and real-time responses via wireless networks, continuous improvement, conversion efficiency, expenses, effluents, and many other crucial landmarks within the business context. The IIoT has an impact on large-scale operations in agribusiness, oil, gasoline, and other industries as well as manufacturing. Similarly, IIoT is having a big impact on the sector thanks to IoT. There are many crucial areas in both IoT and IIoT that guarantee the smooth operation of connected equipment. These enabling technologies can be divided into four major functional categories as depicted in Table 7.1.

TABLE 7.1 Functional Layers of Enabling Technologies

Functional Layers	Enabling Strategy
FL1 Application layer	Industrial innovation
	Framework
	Programs and interfaces
FL2 Middleware layer	Storage service
	Information service
	Memory management
FL3 Networking layer	Message transfer instructions
	Networking link
	Embracing method
FL4 Object layer	Physical entity service
	Entrenched items
	Motorized and electrical components

The FL1 file has three distinct layers. Smart homes, smart cities, and smart agriculture are just a few examples of the initial IoT applications. Next is the architectural layer, which includes supporting software architectures such as REST (Representational State Transfer) and SOA (Service-oriented architecture). The third layer consists of the software and APIs that bridge the gap between the middleware and application worlds. It ensures that all installed programs and operating systems are up to date. Typically, Android or a custom operating system is used to power an IoT device. The implementation of an Internet of Things system may also require specially built application programming interfaces, both of which are critical technologies for bridging the gap between tiers 1 and 2. In the world of middleware, there are three further tiers. The first is virtualization software, which offers methods for obtaining computing resources on demand from remote servers. The three largest providers of cloud services are Google, Amazon, and Microsoft. Secondly, data processing is made feasible by applications of data mining. MongoDB, which provides robust storage engines, is a great illustration of this type of component in an IoT/IA design. Third in the IoT architecture is the cluster field, often known as FL3, which comprises three empowering layers. Layer 1 is dedicated to the protocols used for communication inside a given system; this layer integrates the AL, TL, and NL into a single component. The network interface is the second enabling layer. Important technical guidelines for the IoT infrastructure are provided, paving the way for the system's seamless expansion.

The third and last layer of FL3 is the network layer, which is responsible for adoption processes. Comprised of the gateway layer, connection interfaces, and acceptance layer, these standards are essential to the expansion of the IIoT system since they allow for more reliable wireless communication using technologies like 6TiSCH and IEEE 1095. As can be seen in Table 7.1, FL4 is the last component of the IoT system. The FL4 is composed of three empowering domains. Hardware platforms, which include gadgets like the Arduino and Raspberry Pi, make up the first group. Sensors, radio tags, displays, and firmware are some examples of the last three layers that are brought together in this industry. These parts are crucial for the system to function as a whole. Powering the gadget, in this case, is the domain of the mechanical and electrical components, such as batteries and central processing units. The IoT domain is divided into four functional blocks, making it easier to comprehend the 12 enablers that make up the system as a whole.

7.3.2 BLOCKCHAIN-BASED DIGITAL TWINS

Take into consideration numerous design features associated with DTs, blockchain, and security as you analyze and assess the management studies using a wide range of criteria. An in-depth analysis of reliable blockchain-based DTs seeks to resolve a number of issues raised by previous attempts. Evidence suggests that academics are investigating ways to combine blockchain technology and DTs, showing a growing appetite for DT-blockchain integration. Figure 7.3 illustrates a block diagram of Hardware and Software Interconnections of blockchain.

7.3.3 COMPONENTS OF BLOCKCHAIN

1. **Data Block:** Blocks that make up a blockchain reveal events that have taken place on the platform. The transactional data can be provided as information sharing or as a token transmission procedure that takes place in a system.
2. **Ledger Distribution:** A distributed ledger is a database that is replicated and shared among all nodes in a network. Any user on the network can access this central repository of information.

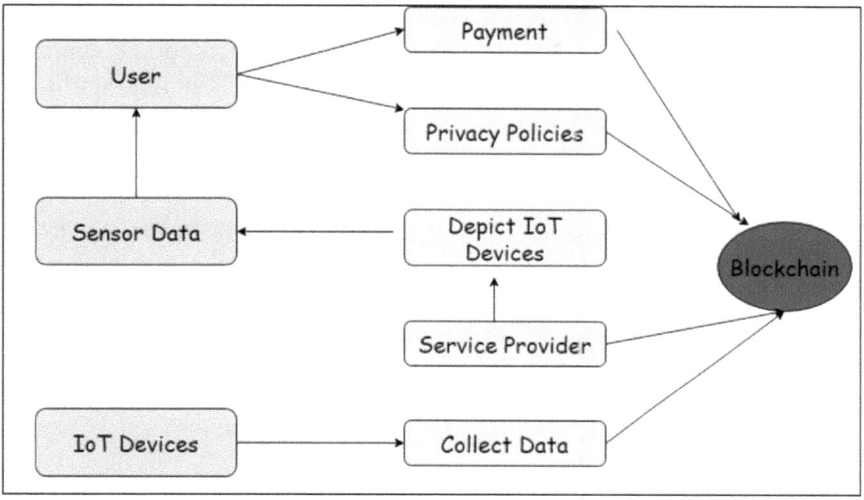

FIGURE 7.3 IIoT blockchain using sensors.

Keeping track of deals is similar to how people in a network exchange information with one another. In a decentralized setting, where no trusted third party is required to monitor and supervise the system's operation, the participating nodes can automatically come to an agreement using a well-established consensus method. Since each entry in the ledger is stamped with a unique cryptographic signature and timestamp, the entire database can be verified and cannot be altered after it has been created. Any alteration to the transaction will produce a new hash inside its branch, which can be easily detected using computational speed and efficiency.

3. **Consensus Algorithm:** In a blockchain system, there are no centralized organizations to control and enforce the standards governing transactions or protect data from security risks. Consensus algorithms seek to safely update the replicated shared states and ensure that at any given time, all replicas of the shared states are synchronized and in agreement. A blockchain's consensus method is a technique for several untrustworthy nodes to decide on a single data log.
4. **Smart Contracts:** To monitor and conduct transactions in line with specified terms and conditions, blockchain applications

known as smart contracts are used. To be more precise, a smart contract is a "digital representation of a temporary economic agreement between two or more parties." Unlike traditional contracts, which rely on centralized authorizing entities to enforce their terms and conditions, smart contracts on a blockchain network do not require such intermediaries to function properly. Smart contracts have become more popular in the blockchain community since the first smart contract platform, Ethereum, was released in 2015. Another name for a smart contract is an "autonomous agent" or a "self-executing engine." As soon as the criteria specified in the contract are satisfied, the code automatically carries out the contract's established provisions, a concept known as "self-execution."

7.3.4 TYPES OF BLOCKCHAINS

Blockchains can be divided into a variety of sorts, each with its unique characteristics, depending on how they are utilized. Generally speaking, there are three types of blockchains: public (or permissionless), private (or permissioned), and consortium (or federated).

1. **Public Blockchain:** Because a public blockchain is an open and accessible network, anyone may join, conduct transactions, and take part in the consensus procedure. It also goes by the name "permissionless blockchain" and operates entirely decentralized and dispersed. Anyone can keep a copy of the blockchain and participate in the process of validating new blocks thanks to the permissionless blockchain. Typically, cryptocurrencies like Bitcoin and Ethereum employ this kind of blockchain technology. As a high number of anonymous nodes can often run on a permissionless blockchain, it is crucial to prevent malicious activity.
2. **Private Blockchain:** On the other hand, a private blockchain is an exclusive network controlled by a single entity. To publish or issue transactions on this blockchain, all parties must first receive permission from a validation process. This suggests that every node joining a private blockchain is a recognized and approved

employee of a single company. A private blockchain is often appropriate for a single enterprise solution and utilized as a distributed synchronized database to track information flows between several offices or people. Particularly, the incentive mechanism is not required for a private blockchain to function, hence a transaction processing charge is often not required.

3. **Consortium Blockchain:** If you're familiar with the settings of a private blockchain, you'll find that a consortium blockchain, also known as a federated blockchain, has a similar permission system. Consortium networks typically encompass more than one company and aid in keeping everyone on the same page. Information transactions between consortium members may be tracked using a consortium blockchain, which acts as an auditable and consistently synced distributed database. Consortium blockchains, like private blockchains, often incur no costs while processing transactions or publishing a new block, both in terms of money and computing resources. The consortium blockchain, in contrast to public and private blockchains, is widely used in massive industrial IoT systems.

Based on the preceding inquiry, Table 7.2 compares various blockchain types. Industrial IoT often employs either a private blockchain or a consortium blockchain due to the confidentiality and safety of manufacturing applications.

TABLE 7.2 Comparison of Various Blockchains

Particular	**Blockchain Types**		
	Public	**Private**	**Consortium**
Involvement	All nodes	Single system	Nodes are chosen in different organizations
Individuality	False anonymous	Valid participants	Valid participants
Admittance	Read/write	Restriction	Restriction
Immutability	True	Partial	Partial
Low permission	True	False	False
Speed of transaction process	Slow	Fast	Fast
Measures of application	Large	Small	Medium

7.3.5 IIoT AND BLOCKCHAIN INTEGRATION

The IoT digitizes and streamlines manual processes, allowing for unprecedented levels of information to be gleaned from the copious volumes of data it generates. This information is paving the way for the development of intelligent apps that will improve city administration and the standard of living for its residents as a result of the widespread digitalization of city services. In recent years, advancements in cloud computing have helped provide the IoT with the tools it needs to analyze, interpret, and translate data into real-time movements and information. There has been extraordinary evolution in IoT, which has opened up several possibilities for people to interact with one another and with information. The principle of open data lies at the heart of these initiatives. As has been the case in a number of previous attempts, however, a lack of trust is a major weakness. The advent of centralized systems, such as the cloud computing concept, has been extremely beneficial to IoT. Data openness is secondary, and network participants are in the dark about what happens to the information they provide.

The benefits of combining IoT with cloud computing have become clear. Furthermore, we acknowledge that blockchain might significantly alter the IoT landscape. The IoT can benefit from blockchain technology because it provides a secure platform for data sharing that ensures the integrity and verifiability of all transactions. Since the data is immutable throughout time, it increases the security of the data and allows the identification of the data sources at any moment.

When many individuals need to exchange IoT data in a secure manner, this incorporation would characterize a significant paradigm shift. For instance, comprehensive tracking of food items is crucial for guaranteeing food safety. Multiple stakeholders, such as those engaged in production, feeding, handling, distribution, etc., may need to be included in food traceability efforts. Better management in these extents might upsurge in protection for food, enhance data sharing among participants, and save search time during a foodborne epidemic, all of which could save lives. Trustworthy data exchange may also boost the services and adoption of ecosystems in other ranges, such as smart cities and smart vehicles. Therefore, using blockchain technology can improve the IoT by providing more reliable and protected data. We believe there has never been an in-depth study of the challenges inherent in the development and rollout of DTs built on top of distributed ledger technology.

Recent research, however, has mostly focused on blockchains' potential in DTs. The findings of the considerably more crucial study classify DTs and shed light on their stages, suggested and implemented iterations, enabling methods, and potential in fundamental needs. Case studies of DTs and collaboration in the automotive sector are shown, and the author discusses the advantages of adopting blockchains for DTs. Scalability issues, data privacy issues, resource-constrained device issues, and interoperability issues brought on by different consensus mechanisms, negotiation systems, and smart agreements are all discussed in this research, along with other technological and societal obstacles to the widespread use of blockchain in DTs. Although it does include security features like confidentiality and integrity that are necessary to safeguard DT's design and operational data, it glosses over other safety-related design aspects like security safeguards and part immunity.

Gains that this interface can provide more precisely comprise (but are not limited to):

- Delegation and extensibility;
- Distinctiveness;
- Self-sufficiency;
- Trustworthiness;
- Retreat;
- Exchange platforms;
- Protected program execution.

One further thing to think about is how the various IoT infrastructures interact with one another. When incorporating blockchain, one must decide whether these interactions will take place within the IoT, on the blockchain, or some combination of the two [35]. A gateway is typically used in an IoT deployment, and it is responsible for relaying sensor data to higher levels. End nodes with low resources are used to communicate with the gateway (cloud or server). To facilitate communication between end nodes and the blockchain, cryptographic functionality may be included in IoT devices. Achieving autonomy in the IoT, which calls for increasingly costly hardware and computing resources, is crucial. Gateway integration is possible with these systems in the same way it is with traditional installations, but the benefits of using blockchain in this way are limited.

Despite blockchain's success, it's still not practical for many users since databases can serve the same purpose with less effort.

7.4 CHALLENGES AND RESEARCH SOLUTIONS

Since both blockchain technology and industrial IoT are still in their early stages, integrating them will present several technological difficulties [36]. The next generation of industrial IoT will be significantly impacted by blockchain, but it will take a lot of work to standardize the designs and standards for both industrial IoT and blockchain. Further investigation is required into the technological, operational, and communal issues of blockchain-based DT schemes to ensure secure data behavior in the Industry 4.0 Era. The outstanding challenges that are specifically hindering the industry's adoption of blockchain-enabled DTs are highlighted in the next section. A comprehensive taxonomy of current and upcoming research and deployment difficulties faced by DTs driven by blockchains is depicted in Figure 7.4.

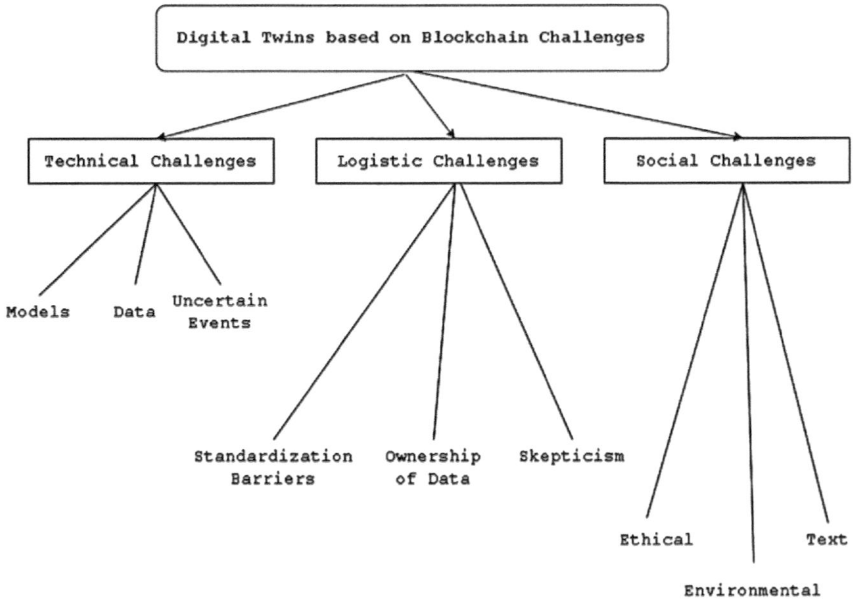

FIGURE 7.4 Various challenges of digital twins.

1. **Multidisciplinary:**
   - The diverse ecosystems utilized for creating and developing present one key obstacle.
   - These occur as a result of the many diverse study disciplines that are involved in collaboration, which can help with breakthroughs but also be a burden because aims in several sectors ultimately steer the research in different directions, slowing down outcomes.
2. **Standardization:**
   - The absence of uniformity, as mentioned in this study and observed in many new and developing technologies, is an additional difficulty for digital twins.
   - As a result, projects in the Digital Twin differ from one another.
   - The range of definitions seen is a contributing element; this, together with the lack of consistency, is a difficulty, impeding the development of digital twin technology.
3. **Global Advancements:**
   - There are several advantages and challenges associated with the huge improvements in digital twins and the technologies that enable them.
   - The adoption and development of newly adapted technology and viewpoints are slowed down by futuristic and unattainable aspirations.
   - The advent of Industry 4.0 and the quickening rise of digitalization bring with them the development of technologies at various rates. However, the rapid expansion creates a number of issues with connectedness and transfer of data.
   - For instance, the secondary systems or Internet of Things gadgets might not work with the Digital Twin. In addition to creating new security difficulties, its growth at various speeds may potentially reveal weaknesses.

7.4.1 RESEARCH SOLUTIONS

7.4.1.1 TRANSACTION SOLUTIONS

- When addressing the increased latencies with conventional blockchain protocols, the session structure is important.

- Interactions in industrial applications must enable user-defined data structures, unlike transactions in cryptocurrencies.
- Transaction size can have a big impact on confirmation delay, especially in IIoT networks with unstable wireless channels.
- Small transactions often have short transmission delays and high transmission success rates. User Datagram Protocol (UDP), a lightweight mechanism, is still widely used in industrial applications as a communication protocol.
- Because of its simplicity, the UDP protocol often provides only the most fundamental forms of communication; for example, it does not provide error correction.
- Lightweight industrial applications require special attention.
- For instance, keeping the transaction size within the limits of network protocols is preferred to limit network frame fragmentation and maximize effective communication (e.g., UDP and IP).
- It's more likely that the participating nodes will mine these reduced transaction sizes.

7.4.1.2 MOTIVATION SYSTEMS

- Offering an incentive is the most common method for encouraging nodes to "work hard" and take part in a consensus mechanism that has already been established.
- In the blockchain, a transaction fee is generally included, and this charge is a critical mechanism for controlling the use of the blockchain's resources and balancing transaction costs.
- To illustrate, transaction costs may reflect the complexity or time sensitivity of a deal. Higher transaction fees are typically charged for transactions that use more resources. Additionally, the transaction fees can offer a way to appropriately redistribute network resources, particularly for public blockchains with restricted capacity.
- When many transactions are generated at once, the time it takes for each transaction to be confirmed may be significantly longer.
- The originator of operations can increase transaction costs to encourage the miner to provide those activities precedence and better public services in order to address this contradictory rivalry.

7.4.1.3 SMART CONTRACTS

- Secured code execution, also known as a smart contract, is a computer-based operation that may be executed independently of a trusted third party.
- Once certain conditions are satisfied, the smart contract becomes active and enacts the relevant contractual requirements.
- The smart contract can operate as a real-time auditor since all transactions in the distributed ledger must be logged and validated.
- A smart contract can either be implemented externally, where the receiver checks the transactions using the smart contract address, or internally, where the code is run.
- Blockchain technology allows for tracking all aspects of performance.
- The smart contract is dynamically executed independently on each node of the blockchain network.

7.5 DISCUSSION

7.5.1 COMMON REQUIREMENTS AND FINDINGS

1. **Data Modeling:** The context of the Digital Twin model and its initial set of findings are discussed. The literature lacks unifying paradigms or a general Digital Twin design, and there is no agreement on the best way to create a Digital Twin system. A general approach to putting into practice a fundamental Digital Twin system is to create a conceptual framework for creating one.
2. **Differentiated Systems:** It is a common objective in the field for academics to obtain a greater grasp of how to handle such systems because the Digital Twin environment is heterogeneous and interconnected to vast telecommunications environments. This can be accomplished by assessing and contrasting existing Digital Twin systems, looking at how they vary in how they handle both smaller and greater diverse Digital Twin settings.
3. **AI:** As digitalization has progressed, artificial intelligence (AI) has become the central component of systems that facilitate

digital twins. The potential effects of AI, ML, and DL algorithms on the growth and use of digital twins are the subject of ongoing research. These advancements in Industry 4.0 principles have enhanced people's quality of life, productivity, and connections as the need for Digital Twin technology grows. It has opened up a wealth of new avenues for study. More academic articles may be written about predictive maintenance and the remaining usable life of equipment in an industrial setting now that AI is available to do these tasks.

4. **Security:** With blockchain technology's advancements, quite a lot of options to maintain retreat, with a primary emphasis on investigating elucidations that support safe Digital Twins.
5. **Exchange of Data:** Similar to security and AI, the emergence of Digital Twin systems and their supporting infrastructure coincides with networking. Further investigation will ensure that the Digital Twin can thrive when data is shared among devices. Finding methods to achieve seamless data integration for both small IoT units and wide and diverse systems is an ongoing topic. It is necessary to scale up the creation of small digital twins while considering limited data sharing.

7.6 CONCLUSION

The integration of blockchain and IIoT platforms are thoroughly and methodically reviewed in this chapter. This chapter discusses some of the main problems and challenges with this connection between blockchain and IIoT platforms. The potential drawbacks of this integration are then discussed, along with a few important industrial application areas. Finally, several prospective paths for further research on the fusion of blockchain and IIoT platforms are provided. Ensuring the reliability of data in the IIoT is a challenging task that requires the implementation of crucial controls at every stage of data production and dissemination. By comparing prevailing blockchain-based DTs, research fields in design solutions involving DTs, blockchain, and security were identified. The various findings of the chapter help guide academics through obstacles and lay the foundation for the adaptation of blockchain-based DTs in engineering applications.

Although the concept of the digital twin is still new and heavily influenced by the industrial sector, the current analysis sets the stage for further research in this area.

KEYWORDS

- **blockchain**
- **data analysis**
- **data integration**
- **digital twin**
- **industrial internet of things**
- **machine learning models**
- **spatial networks**

REFERENCES

1. Xu, X., Lu, Y., Vogel-Heuser, B., & Wang, L. (2021). Industry 4.0 and Industry 5.0—Inception, conception and perception. *Journal of Manufacturing Systems, 61*, 530–535.
2. Prasanth, A., & Jayachitra, S. (2020). A novel multi-objective optimization strategy for enhancing quality of service in IoT-enabled WSN applications. *Peer-to-Peer Networking and Applications, 13*, 1905–1920.
3. Jamil, S., Rahman, M., & Fawad, A. (2022). Comprehensive survey of digital twins and federated learning for the industrial internet of things (IIoT), internet of vehicles (IoV), and internet of drones (IoD). *Applied System Innovation, 5*, 56.
4. Fei, T., Yongping, Z., Ying, C., & Jiawei, R. (2022). Digital twin and blockchain enhanced smart manufacturing service collaboration and management. *Journal of Manufacturing Systems, 62*, 903–914.
5. Liu, Y. K., Ong, S. K., & Nee, A. Y. C. (2022). State-of-the-art survey on digital twin implementations. *Advanced Manufacturing, 10*, 1–23.
6. Ghobakhloo, M. (2020). Industry 4.0, digitization, and opportunities for sustainability. *Journal of Cleaner Production, 252*, 119869.
7. Chen, M., Herrera, F., & Hwang, K. (2018). Cognitive computing: Architecture, technologies and intelligent applications. *IEEE Access, 6*, 19774–19783.
8. Jazdi, N. (2014). Cyber-physical systems in the context of Industry 4.0. In *Proceedings of the 2014 IEEE International Conference on Automation, Quality and Testing, Robotics* (pp. 1–4). IEEE.

9. Boyes, H., Hallaq, B., Cunningham, J., & Watson, T. (2018). The industrial Internet of Things (IIoT): An analysis framework. *Computers in Industry, 101*, 1–12.
10. Dang, L. M., Piran, M. J., Han, D., & Min, K. (2019). A survey on the Internet of Things and cloud computing for healthcare. *Electronics, 8*, 768.
11. Kavitha, M., & Roobini, S. (2023). Systematic view and impact of artificial intelligence in smart healthcare systems, principles, challenges, and applications. In *Machine Learning and Artificial Intelligence in Healthcare Systems* (pp. 25–56).
12. Bilberg, A., & Malik, A. A. (2019). Digital twin-driven human-robot collaborative assembly. *CIRP Annals, 68*(1), 499–502.
13. Thang, P., Trang, T., & Tai, B. (2022). Multimedia privacy, security, and protection within the blockchain: A review. In *Proceedings of the 5th IEEE International Conference on Contemporary Computing and Informatics* (pp. 1171–1176).
14. Mohammadi, N., & Taylor, J. E. (2017). Smart city digital twins. In *Proceedings of the IEEE Symposium Series on Computational Intelligence (SSCI)* (pp. 1–5).
15. Chao, Z., Guanghui, Z., Han, L., & Yan, C. (2020). Manufacturing Blockchain of Things for the configuration of a data-and knowledge-driven digital twin manufacturing cell. *IEEE Internet of Things Journal, 7*(12), 11884–11894.
16. Raj, P. (2021). Empowering digital twins with blockchain. In *Blockchain Technology for Secure and Smart Applications in a Variety of Industries* (pp. 267–283). Elsevier's Advances in Computers, 121.
17. Götz, C. S., Karlsson, P., & Yitmen, I. (2022). Exploring applicability, interoperability, and integrability of blockchain-based digital twins for asset life cycle management. *Smart and Sustainable Built Environment, 11*(3), 532–558.
18. Jiewu, L., Guolei, R., Pingyu, J., & Kailin, X. (2020). A survey of blockchain-enabled sustainable manufacturing and product lifecycle management in Industry 4.0. *Renewable and Sustainable Energy Reviews, 132*, 110112.
19. Trond, K., Adil, R., & Omer, S. (2020). From a modeling standpoint, what are the benefits, constraints, and facilitators of a digital twin? *IEEE Access, 8*, 21980–22012.
20. Ross, D. (2016). Digital twinning information technology virtual reality. *Engineering & Technology, 11*, 44–45.
21. Laaki, H., Miche, Y., & Tammi, K. (2019). Prototyping a digital twin for real-time remote control over mobile networks: Application of remote surgery. *IEEE Access, 7*, 20325–20336.
22. Mohammadi, N., & Taylor, J. E. (2017). Smart city digital twins. In *Proceedings of the IEEE Symposium Series on Computational Intelligence (SSCI)* (pp. 1–5).
23. Pargmann, H., Euhausen, D., & Faber, R. (2018). Intelligent big data processing for wind farm monitoring and analysis based on cloud-technologies and digital twins: A quantitative approach. In *Proceedings of the IEEE 3rd International Conference on Cloud Computing and Big Data Analysis (ICCCBDA)* (pp. 233–237).
24. Ruohomäki, T., Airaksinen, E., Huuska, P., & Kesäniemi, O. (2018). Smart city platform enabling digital twin. In *Proceedings of the International Conference on Intelligent Systems (IS)* (pp. 155–161).
25. Sivalingam, K., Sepulveda, M., Spring, M., & Davies, M. (2018). A review and methodology development for remaining useful life prediction of offshore fixed and floating wind turbine power converter with digital twin technology perspective. In

Proceedings of the 2nd International Conference on Green Energy and Applications (ICGEA) (pp. 197–204).
26. Qi, Q., & Tao, F. (2018). Digital twin and big data towards smart manufacturing and Industry 4.0: a 360-degree comparison. *IEEE Access, 6*, 3585–3593.
27. Kuehn, W. (2019). Simulation in digital enterprises. In *Proceedings of the 11th International Conference on Computer Modeling and Simulation (ICCMS)* (pp. 55–59).
28. Xu, Y., Sun, Y., Liu, X., & Zheng, Y. (2019). A digital-twin-assisted fault diagnosis using deep transfer learning. *IEEE Access, 7*, 19990–19999.
29. Chhetri, S. R., Faezi, S., Canedo, A., & Faruque, M. A. A. (2019). QUILT: Quality inference from living digital twins in IoT-enabled manufacturing systems. In *Proceedings of the International Conference on Internet of Things Design and Implementation* (pp. 237–248).
30. Lu, Y., Peng, T., & Xu, X. (2019). Energy-efficient cyber-physical production network: Architecture and technologies. *Computers & Industrial Engineering, 129*, 56–66.
31. Umeda, Y., Ota, J., Kojima, F., & Saito, M. (2019). Development of an education program for digital manufacturing system engineers based on the 'digital triplet' concept. *Procedia Manufacturing, 31*, 363–369.
32. Kobara, K. (2016). Cyber-physical security for industrial control systems and IoT. *IEICE Transactions on Information and Systems, 4*, 787–795.
33. Fei, X., Shah, N., Verba, N., & Chao, K. M. (2019). CPS data streams analytics based on machine learning for cloud and fog computing: A survey. *Future Generation Computer Systems, 90*, 435–450.
34. Čolaković, A., & Hadžialić, M. (2018). Internet of Things (IoT): A review of enabling technologies, challenges, and open research issues. *Computer Networks, 144*, 17–39.
35. Bhaskar, K. B., Prasanth, A., & Saranya, P. (2022). An energy-efficient blockchain approach for secure communication in IoT-enabled electric vehicles. *International Journal of Communication Systems, 35*, 1–20.
36. Dhanaraj, R., Kadry, S., & Kang, B. (2022). Probit cryptographic blockchain for secure data transmission in intelligent transportation systems. *Journal of Internet Technology, 23*, 1303–1313.

CHAPTER 8

Leveraging IoT and AI Technologies for Blockchain-Based Digital Twins

S. ROOBINI, M. S. KAVITHA, and S. KARTHIK

Department of Computer Science and Engineering, SNS College of Technology, Coimbatore, Tamil Nadu, India

ABSTRACT

The Industrial Revolution is being driven by new technologies like digital twins, blockchain, the Internet of Things (IoT), and Artificial Intelligence (AI), which are also having a significant impact on the healthcare industry. Innovative technologies like blockchain, IoT, and AI will help bring about digital transformation and upend many businesses. As these three technologies converge, new business models will emerge: Sensors, automobiles, machines, and other autonomous agents will function as their profit centers, sending and receiving money autonomously using blockchain technology, and making choices independently as autonomous financial entities employing analytics and machine intelligence. The creation of standalone business strategies and the role of technology will be fueled by this confluence of knowledge. The physical data must be cross-verified using many device units with corresponding points of vision in order for it to be reliable. The blockchain may therefore be used to store cross-validated data by means to preserve integrity and the authenticity of the information. By utilizing AI and IoT technologies, as well as by spotting errors and suggesting preventative steps in advance of crucial occurrences, Digital Twins (DTs) may be used in this respect to derive clever interpretations of the facts. The primary issues of dispersed intelligence warehouses

Blockchain-Based Digital Twins: Research Trends and Challenges.
A. Prasanth, Vandana Sharma, Rajesh Kumar Dhanaraj, Pethuru Raj (Eds.)

are unreliable fact distribution and the requirement for forecasting management, which are targeted through DT empowerment commercial utilization for blockchain. This chapter provides a thorough analysis of the most recent research findings for blockchain-based digital twins while highlighting the major advantages of employing DTs. These emphasize the part AI plays in DTs built on blockchain technology. It also mentions prospective study and deployment challenges with technology facilitated by blockchain that need to be looked at additionally.

8.1 INTRODUCTION

The IoT, AI, DT, and Blockchain technologies are now acknowledged as advancements with the potential to disrupt whole sectors and enhance existing commercial procedures [1]. As a result of offering a unified and highly consensus ledger in which blockchain can improve enterprise disclosure, stability, and privacy operations. Similar to a register, a shared inventory, or typically a network may hold any form of asset. This information mostly relates to money and identity. IoT is the driving force behind user-friendly business operations and industry-essential automation. In addition, AI optimizes the results of various business processes by identifying trends. Until now, the connections between these advancements have frequently been overlooked by blockchain, IoT, and AI. But in the future, these ideas will merge and should be used together. The purpose of the Internet of Things is to gather and distribute solutions that enable the framework and facts and also establish the criteria of interaction. AI improves guidelines and procedures may be one way these technologies are related. These three technologies complement one another by design and when integrated, they may reach their full potential.

Blockchain technology may bring about unforeseen industrial and economic disruption taking place in a universal range [2]. A blockchain is a distributed ledger that contains links flanked by individual blocks of a digital transaction. The differences between blockchain technology and conventional internet technologies are numerous and substantial. On the internet, only data and copies of items are transferred, and the original data is not preserved. Its significance is reflected within a time-stamped event securely maintained in an online record through the use of a blockchain. Blockchains serve as a digital ledger for cryptocurrency transactions; furthermore, innovation is predicted to completely transform the global electronic financial system. Additionally, by employing multiple trials and real-world examples of operations to generate DTs, enhanced medical device manufacturers have the potential to have a direct impact on healthcare systems. It is possible to match actual settings with virtual ones by integrating industry, AI, IoT, modeling, and data analytics [3]. DTs are expected to be embraced by industrial organizations and emerge as the dominant technology for the smart industry. Electronic twins can be used in conjunction with the autonomous and distributed database architecture known as blockchain to achieve security, privacy, and history tracking. The majority of smart contract activities include simple techniques that are consolidated to expose hazards that are similar to isolated crash points.

DTs were developed in the manufacturing sector, and are crucial for managing the numerous gadgets that are used in that sector, and for performing user tasks. Through the virtualization of the actual world and the reflection of the outcomes of improved simulations, DTs can be helpful. Data connected to city administration is growing as smart cities are built using robots, automated transportation infrastructure, surveillance, automated vehicles, and the Internet. Decision-making and data processing are both made feasible by the advancement using networks of wireless communication such as 5G, the Internet, cloud-based services, and machine intelligence. With the aid of numerous sensors, these technological advancements have made it feasible to swiftly incorporate real-world structures and occurrences in the virtual environment. DTs for land management are a major global problem. To standardize DTs data, a global organization for the standardization of geographical information is exerting major effort. Information Automation such as big data, Intelligence, and IoT become more and more important over time.

8.1.1 CHALLENGES

The adoption of DTs cooperation in decentralized systems still faces a number of difficulties since the electronic twin paradigm is currently being developed. These difficulties include the following:

1. **Interoperability:** Determining the data format for DTs and the need for cooperation must describe modeling as well as methods of expressing regulations.

2. **Authentication:** In other cases, independent entities who desire to work together on the deployed DTs in decentralized systems. Therefore, in order to gain authentic, protected information interchange and evaluation among various participants, smart contract devices require innovation that is reliable and effective.
3. **Distributed Machine Learning:** To make precise predictions about the possible dangers in the system, a vast amount of input data from several individuals must be examined.
4. **Data with a Single Failure:** Affect centralization, whereas data with no single failure affect decentralization.
5. **Scalability and Robustness:** In decentralized systems, a system must support a high number of DTs that represent various players, such as items, machines, workstations, nodes, and so on. Additionally, the decentralized system must manage several distributed DTs while simultaneously maintaining robustness at a needed level, particularly in situations of hacked nodes and malfunctioning circumstances.

In order to provide mentor transmission, independent file storage, accessibility, safe sustainability, and provenance, the majority of research projects have suggested the implementation of blockchain using smart technology. The current and next difficulties in research will be a thorough analysis of the most recent findings for blockchain-based DTs. Additionally, by collaborating hierarchically and finely leveraging common knowledge to control and then track the product assembly data, numerous blockchain-based DTs may be used. Furthermore, a smart contract is employed to carry out various tasks automatically in order to effectively boost data exchange and offer greater security. It offers trustworthy data provenance, audit, traceability, and transaction monitoring for players in the development of DTs.

8.1.2 SCOPE

An exhaustive, systematic study of cutting-edge outcomes for blockchain-based DTs, a thorough examination of the planning and execution problems with cryptocurrency-based DTs, and provision with potential fixes

[4]. Additionally, look at any technical, logistical, and social challenges that emerge as a result of integrating blockchain with DTs and that stakeholders need to deal with.

8.1.2.1 CONTRIBUTIONS

The chapter's key contributions are.

- The chapter comprises a structure of DTs along with the association of blockchain technologies, and categorization of blockchain-based DTs and their potential integration.
- Designing the solutions for DTs and blockchain in addition to lifecycle, has several properties, an authentic framework, and security mechanisms with principal advantages.
- Integrating Intelligence and IoT along with blockchain and DTs invokes challenges with deployment regarding scientific and logical issues.

8.2 CONCEPTUALIZATION OF IOT, AI, DT, AND BLOCKCHAIN

The evolution of DTs and blockchain is supported by a number of critical technologies considering the Internet of devices and AI in this segment.

8.2.1 DIGITAL TWINS

With their data, functions, and communication capacities, DTs represent tangible items or persons in the electronic realm. The term DT refers to a digital model that resembles an entity's real-world behavior based on data. To be very realistic with actual information, the virtual model employs electronic modeling and insights collected from deep fundamental and ecological monitoring and assesses the characteristics of a physical entity. The fundamental distinction between a DT and an online scenario is how a digital twin is created, which depends on a subject's personal traits, remote monitoring, and current information.

The device known as DTs replicates physical things in the virtual realm and operates on the same principles as it does in the physical one. Because it embodies them like twins, the fact of the online environment is referred to as electronic twins. Initial DTs replicated the artificial world's realism, yet current advancements in IoT, computer, and AI technologies have made it possible to incorporate a variety of scenarios that are not possible to imitate owing to practical constraints. Realistic forecasts may be made, and issues could be addressed by employing a variety of detectors that provide actual data. It is a DT platform that, using simulation based on the gathered data, operates real-world equipment more effectively. The structure of a DT is exposed in Figure 8.1. It employs model data and data from actual sensors to transform real items into digital models, which it then uses to recreate actual objects through data analysis, prediction, and visualization [5].

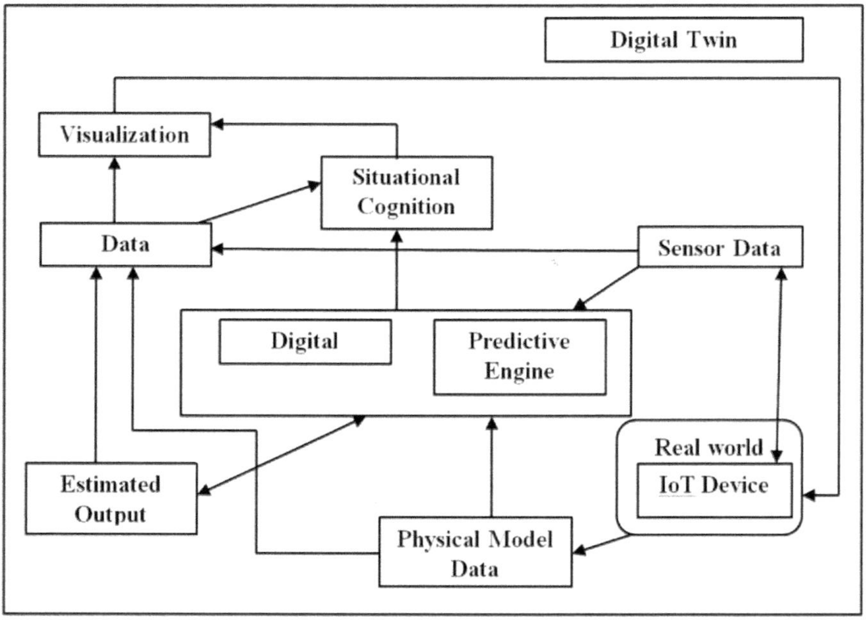

FIGURE 8.1 Framework of DT.

Due to their potential to help in the modeling, forecasting, and optimization of a variety of digital artifacts, DTs are being explored and implemented at an increasing rate. Some of the main issues with using digital documents and data include safety and confidentiality, as well as the dependability of information sources. Determining the advantages that a DT might offer at the design stage is crucial in the recent automation method. Following are some advantages of using business information management and DTs with blockchain in the direction of the building industry:

1. **Global Administration and Surveillance in Real-Time:** The simulated companion would provide a thorough understanding of digital assets that can be directly managed by regulatory mechanisms. With a relational network, all operations may be captured together with data changes, enabling real-time valuable intelligence to be used to make informed decisions.
2. **Improved Security and Efficiency:** When devices are controlled remotely, computers and clients are able to carry out creative virtual models that might enhance protection and regulation by allowing employees to operate in a trustworthy environment.
3. **Planning and Maintenance Estimation:** For extended electronic matching, several sensor monitoring facilities may provide significant amounts of information in real-time. Problems with the equipment can be discovered considerably sooner owing to intelligent results analysis and performance monitoring. These goals aim to minimize time by providing faster maintenance and certification performance.
4. **Instance and Threat Assessment:** With a DT, what-if analysis is feasible, allowing for better risk management. In order to deal with unforeseen occurrences, such analysis might evaluate the system response along with additional compensatory techniques.
5. **Cooperation:** Using BIM and Blockchain, teams can strengthen their affinities and coalitions through interaction, resulting in increased production, more autonomy, and the avoidance of delays while eliminating middlemen.
6. **Distributed Data Operations:** With additional layers for verification and recombination of various digital artifacts, blockchain enables monitoring of pattern modifications and authenticity of prototype data.
7. **Improved Documentation and Transparency:** On a blockchain network, authorities and clients can view transactions, but they

cannot edit data that has already been saved. Real-time information can help disciplines maintain an updated digital replica when combined with automated monitoring to ensure fairness.

8. **More Accurate Statistics and Forecasting:** Access to real forecasting data and enabled platforms may facilitate decision-making for a variety of building component tasks.
9. **Knowledge Tracking:** Blockchain creates a structure for monitoring file identities among systems, enabling tracing in the development and evaluation phase inside a dimensional control. Due to Blockchain's immutability and open nature, security breaches are rare. When using network duplication, a copy of the blockchain system is stored on each node in the system. The Blockchain Network ensures that no files are altered when they are moved by linking software applications as gateway blocks.
10. **Decentralization:** Based on a centralized method, authorities, and individuals may have the ability to connect to the public blockchain that saves secret keys used to share information. A group of connections manages the network, removing limitations imposed by a centralized authority.
11. **Data Control by Several Parties:** Accessibility to the Blockchain Platform is available for development professionals to check the system record and relevant information beyond the capacity to track numerous process iterations.
12. **Technology:** The entire building process is managed automatically thanks to the conversion of contract identity algorithms into payment systems. This guarantees that operations are finished in a timely manner, preventing any bottlenecks.

8.2.2 BLOCKCHAIN

A repository that serves as the foundation of a BIM workflow frequently reveals privacy and accessibility problems, including data breaches that reveal serious threats to data protection, authenticity, and consistency [6]. Blockchain is a dependable development solution that, by supporting information management, provides unambiguous and exact dissemination of information among stakeholders in order to mitigate these hazards. The Distributed Ledger Technology, popularly known as Blockchain, includes values ranging from recording and management systems. Stakeholders in a blockchain environment can confirm and endorse the independence while modifying the data in the nodes.

Model specification, adaptation, and mistake correction are a few of the constraints and issues that the fusion of network technology and enterprise management systems can solve for a DT. By specifying a variety of parameters, including location, tasks, and activity, the foundation of the consensus mechanism is a piece of executing computer code connecting nodes which increases security for implementing the network relies on Blockchain application. They also enable code execution. Blockchain can enhance the development of a DT for many development situations by enabling users of BIM to collaborate and transfer files while somehow offering increased accountability and technical standards. Due to its high substantial requirements for cases utilizing mining, Blockchain is governed by environmental and sustainability rules. Blockchain uses web accessibility account information and an authorization framework, which also poses confidentiality issues, to operate on data.

Transactions between stations serve as the fundamental building blocks of communication in the blockchain and are stored as incremental backup moment records in a permission database. A planned and systematic way that requires tackling an asset-challenging and convenient mathematical issue is used to establish a safe, trustworthy connection among entrusted nodes. Following the acceptance procedure, a block of transactions is put together and added to the registry. Each module consists of the hash in its preceding segment, which promotes authenticity into scalability topology and makes it easy to spot tampering with a block and consequently events.

The development of cryptocurrency has played a major role in the fast deployment of blockchain, which has effectively replaced traditional financial transaction systems in many enterprises. The financial and e-commerce industries are showing to be a very dominant innovation as they continue to develop [7]. Furthermore, a wide range of non-financial application situations are noted as prospective candidates for blockchain-based solutions, including, among others, medical, elections, proprietary information, and verification of position. Recently, IoT and blockchain-based technologies have been growing rapidly to reimagine diverse business structures.

8.2.2.1 TECHNOLOGIES OF BLOCKCHAIN

A blockchain is a continuously growing collection of data connected by cryptography, where each block includes itemized receipts, a timestamp, and a computational hashing of the block before it. Network records are maintained in a decentralized system that can't be altered until all following blocks have been updated.

1. **People-to-People Networks and Blockchain:** DTs are created based on the technology's underlying blockchain. The blockchain was initially suggested as a shared database for currencies to create a continuing network hashing and build immutable data records via a people-to-people network [8]. The people-to-people network is autonomous, and each user, referred to as a node, is in charge of giving the network the required services. Before being stored on the distributed ledger, the block must be verified using confidence processes and reflected in the modified block on the updated register. Because of this, all networks are able to obtain information besides distributing and giving information to other components.
2. **The Blockchain Technology Uses Hash Algorithm:** Chains of connected blocks make up the blockchain, which is made up of individual blocks. This framework component in the blockchain technology comprises a connector and structure in favor of every frame, significantly the header checksum in a block, log file besides locating a period to design a network, the disorganized underlying cause checksum for the preceding frame in accordance with the connectivity rules, and the structure containing encrypted and parsed crucial operations. The component may be present in several events. Network checksums are distinct; therefore, if any changes are made to any block header, the associated checksum of the block will also be changed. One technique used to verify cryptographic reliability is evidence of work.

8.2.3 ASSOCIATION OF BLOCKCHAIN, IOT, DT, AND AI

Blockchain technology up until a few years ago was only mentioned in relation to transactions. Ever more new research highlights the advantages of integrating blockchain – based on certain other breakthroughs like IoT and AI. The application of blockchain-based to enhances different IoT equipment's system architecture. Blockchain design may be changed such that the resultant technology, especially in terms of process efficiency, is increasingly suited to support IoT applications [9]. Some research emphasizes incorporating information with AI along with the relationship connecting technology with IoT.

Currently, rather than using all these technologies at once, the emphasis is mostly on integrating blockchain in tandem with cutting-edge innovation that spans AI and IoT. Furthermore, integrating these advances will be the only way to fully utilize the promise of these novel, developing technologies. Create an IoT and AI-friendly blockchain architecture. A brief explanation of the advantages of every invention but also how they work best together. A real-world application scenario serves as an example of how blockchain technology, IoT, and AI may be merged. The main distinction among the two kinds of blockchain technologies is that when using a public ledger, all users have access to the information within. On the other hand, the Permission blockchain will limit those who may use the information. Yet, whether the accessibility is general or restricted is irrelevant because both blockchain technologies might be used to accomplish the applications.

8.2.4 POTENTIAL OF INTEGRATION

The ability to save data from DTs and communicate with blockchain is a chief advantage of employing electronic twins by means of this technology. DTs can engage in blockchain networks with other related technologies to create blockchain technology that is built on interaction. This first demands that the data related to DTs be unchangeable and then using blockchain is crucial for protecting them. In order to confirm the data source, one option is to store data integration content in the blockchain rather than retaining it in a conventional system. The requirement for DT interaction underscores the significance of merging blockchain with DTs. So, the two core elements of blockchain technology are the immutable network and the consensus mechanism. This consensus mechanism should thus be executed in DTs. Additionally, machine learning may be employed to forecast and determine if events occurred in the automated systems. The advantages of fusing DTs, blockchain, AI, and IoT are highlighted by the following traits:

1. **Data Protection:** Due to blockchain's global nature, it is more difficult for intruders to attack sensitive information since they would need to target every reverse proxy, which is nearly impossible. Additionally, any system or product behaving strangely is promptly removed from the blockchain, ensuring the system's ongoing reliability.
2. **Data Reliability:** Blockchain safeguards information confidentiality through robust authentication and protection. It provides enhanced visibility by monitoring transactions.
3. **Real Information Insights:** Business intelligence offers detailed data analysis, while blockchain technology enables authentic operations. By combining these two techniques, relevant data insights can be generated to streamline corporate operations and revolutionize various sectors.
4. **Forecasts:** Blockchain data analysis can uncover significant data discoveries and patterns within the data.
5. **Data Exchange:** Organizations can avoid reusing previously used knowledge or duplicating completed data analyses by backing up data on a blockchain platform. This approach also facilitates secure data transmission without the need for frequent data editing.

Blockchain and computational modeling, each with their advantages and disadvantages, can complement each other effectively in managing data volume and quality. As blockchain technology advances and evolves, additional use cases such as data analytics will be explored [10]. Conversely, business intelligence can benefit from blockchain's cost-effectiveness. It will be intriguing to observe how these technologies progress in addressing current challenges and how they may transform information systems and utilization.

8.3 BLOCKCHAIN-BASED DTS

Using new blockchain-based digital technology has attracted plenty more, particularly in relation to digital currencies. In blockchain networks, processors employ operations to look for consensus mechanisms or make use of those that already exist. Because there is a maximum number of transactions in each batch, producers must emphasize choosing the best contract, even though they lack sufficient knowledge to do so and can only determine the agreement's prize fund. In order to improve the decision-making process, decision-makers must support innovative ideas to handle the uncertainty in the blockchain environment. The Twin, which is connected to industry and the IoT, gathers data from distant gadgets such as sensors, object models, etc., in order to improve the preference pane and provide the best outcomes in manufacturing plants.

Anxiety considerations affect DT production. Additionally, gives a thorough explanation of a trustworthy electronic-based blockchain that aims towards a few of the unresolved issues within the prior attempts. The typology developed focused on industrial design ideas for blockchain-based smart contracts is shown in Figure 8.2. It begins by discussing design details such as definition, component integration and configurability, operational stages, and complexity of the task. In addition, it explains the kind of network used in the suggested proposal, development considerations, an asynchronous backup and recovery method, accountability, and efficient information providers. Furthermore, it draws attention to designers linked to protection, such as authentication, security systems, and physical robustness [11].

1. **Creating DTs: Solutions:** DTs are viewed as approximations of just about any resource, service, equipment, or production process in the bulk of already extant works. The connection and accessibility of different aspects or sub-sectors offer an accumulated picture of the various structural environments. Nevertheless, they are not encompassed by the previous approaches. Creating a Digital Transformation Agency by merging the token identifiers of the tangible asset and this behavior will provide a wider context of the external reality. As a consequence, conceptual perceptions of more complicated processes are inadequate because of the absence of linkage across several simple DTs.

   The conceptual designs for DTs are covered in the sections that follow from a variety of angles, including the development lifespan, integration, and compatibility, reliability, and stability, and synchronicity and consistency. With the aid of acquired functioning patterns of practices, DTs assess the operational surroundings to see the present state of the equipment and forecast

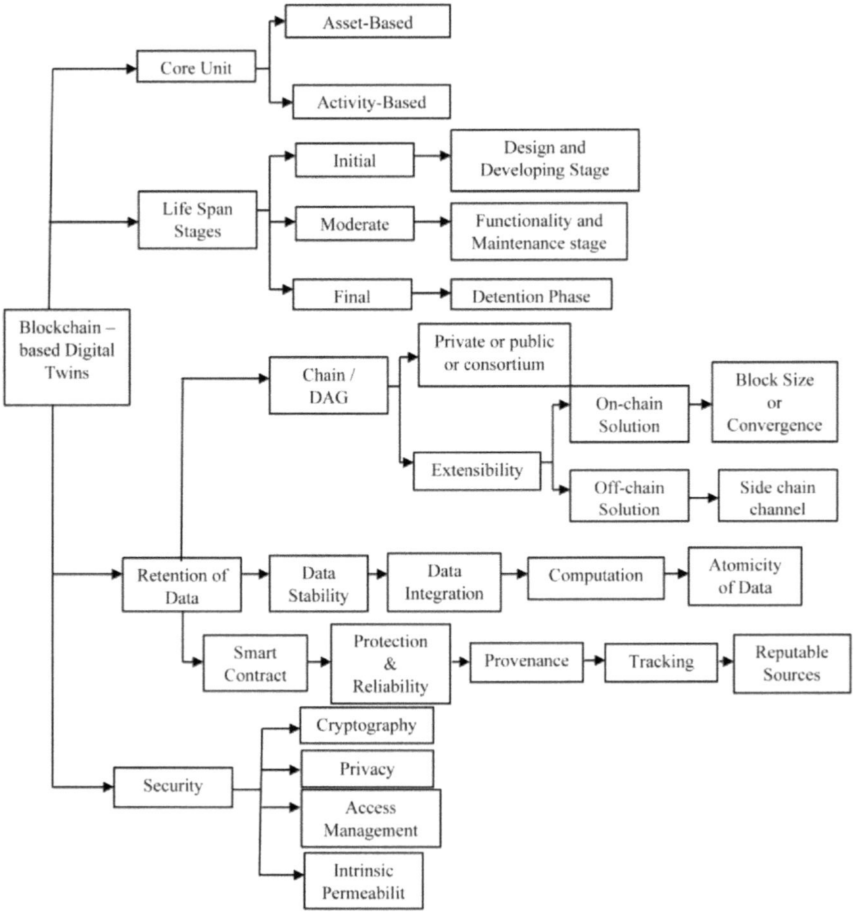

FIGURE 8.2 Classification of blockchain-based DTs.

trends. Each DT offers a unique benefit for product development processes. In relation to Industry, AI, Intelligent machines, and the attainment of continuous DT development, this subsection demonstrates the key influences, dangers, and obstacles that impact the manufacturing and fabrication of DT.

2. **Data Integration and Computation:** One of the key requirements for the deployment of DTs is the integration and compatibility of information from multidimensional and disparate devices. This entails removing inaccurate, redundant, or incomplete information,

integrating distributed information standards into an individual data framework, and finally sanitizing the combined data to produce a coherent representation of a particular item. As demonstrated, the previous approaches take the information hybrid material into account from a technological standpoint. Compatibility is also barely slightly handled. Entrants of networking equipment were established, which allows for the analysis of information gathered from equipment using various network standards. In addition, it has peripheral bridge processors that analyze and consolidate the observed information before evaluating it.

The blending of information in many forms to provide information multiplicity is analogous. They divided the information into categorized information (datasets), unformatted, and semi-indistinct information to depict tangible resources and deterministic attributes of the system. The aforementioned ideas, however, require concrete achievement of tremendous results that would show how practicable they are. In comparison to a small information perspective, connectivity, and data transformation provide a description that is more thorough, coherent, and reliable. Furthermore, the previous efforts merely don't provide the connection involving information synthesis and DT performance with enough consideration.

3. **Protection and Reliability:** To identify failures and to lessen the effects of broken elements and suspicious attacks in a manufacturing operation, internal functional factors, such as stability and protection risks, must be incorporated. Failure to implement such regulations has devastating effects on the network in the presence of assaults or Lofts. Nevertheless, it is a crucial safety technology architectural condition. To the best of our knowledge, just one of the currently scholarly papers issued a noteworthy work-based recommendation for embracing security and safety projects. This work focused on the modification of a vehicle's power and proved the recognition of the infamous intermediary that spies on communication assault attacks [12].

   Upcoming possibilities are also suggested to further examine this parameter, including attitude tracking, assessment combined with intrusion detection systems, examining gadget designs with privacy laws, and programming terminology for safety and security legislation. Later phases of the lifespan are possibly less susceptible to mistakes and accidents as a consequence of the inclusion of protection mechanisms during the initial concept stage. However, in order to guarantee the accuracy of security and safety regulations, the register must be used to maintain and access them. This supports the case for algorithms and blockchain connectivity even more.

4. **Stability and Coordination:** The coordination and uniformity of information aligning the electronic and virtual environment is another crucial aspect of DT architecture. The amount, production attached with storage is significantly correlated with both of these parameters. The two types of DT data are cognitive data and analytical data. The latter varies regularly with each phase of the process, producing distinct forms and a real-time data framework, whereas the alterations seldom occur over time during its lifetime of reality equivalent.

   Issue of accommodating storage capacity and throughput is somewhat taken into account in the present efforts. They recommended using predefined durations to preserve sensor information. Unfortunately, they failed to take into account the crucial need for real information alignment. On the contrary, a more practical method of keeping the cipher connection to the digital signal upon that record while off-chain maintaining the contents was developed. The problems with data storage are handled by this method, although operational criteria information is indeed lacking. Only non-nominal or consolidated collected information should be distributed in order to decrease storage capacity, while the off-chain backup and recovery structure were recommended. A cross mechanism is proposed DT paradigm that refers to a process crucial for living standard management and commercial pro-government using smart contracts as a searchable agent. Additionally, they leverage conventional computers to quickly locate financial records.

   The main issue, when looking at the current methods, is the fact that active information obtained by the installed controllers and devices, which offer direct information on the portfolio's present system, often assembles in morphologically uniting decades and moments. The pace and amount of sensor digital data outperforms the technical specifications of popular blockchain systems [13]. However, choosing whose content to preserve chain-based twins while deciding which to put chain shutdown requires careful consideration. The smart contracts DT technologies must have high performance and low communication frequency in order to successfully interconnect facts across online and offline space, by enhancing information acquisition by utilizing it wisely to handle computer database difficulties.

5. **Atomicity of Data:** Information is regarded as the engine that guarantees exact judgment, yet accuracy of the evidence is just as important for making selections as abundance. As a result, the volume of data gathered remains a crucial process that is designed. The ideal size and positioning of devices, which varies on the level of generality of the data obtained, are yet not addressed in the current studies.

8.3.1 DTS LIFESPAN

A DT instance in which a digital thread may be utilized achieves information connection during all various phases of the lifespan. The DT instance phases begin with design and go all the way through dismissal, without including any potential series of stages parities. Then, during the planning process, the resource model is created and developed based on statistics, dynamic data, and numerical simulations. Figure 8.3 depicts the phases of DTs. During the design phase, a practical resource is then modified according to the resource model and many other design specifications.

The asset continues to improve and enhance its functions throughout the prolongation and functional phase as the model develops and consciousness to the resource throughout its lifecycle, which is the right phase in a coordinated and ongoing relationship amid the sample and the resource. Eventually, in the termination phase, the resource and its DTs are destroyed owing to depreciation and such factors [14]. The initial, intermediate, or both of these phases of the DTs lifecycle have often been the focus of prior research. The specifics of how the item is dismantled and the archival information captured about the DT goods are preserved and made accessible to other items or knowledge workers to enhance the manufacturing of new versions are not provided in the hypothetical blockchain-based DT solution being built.

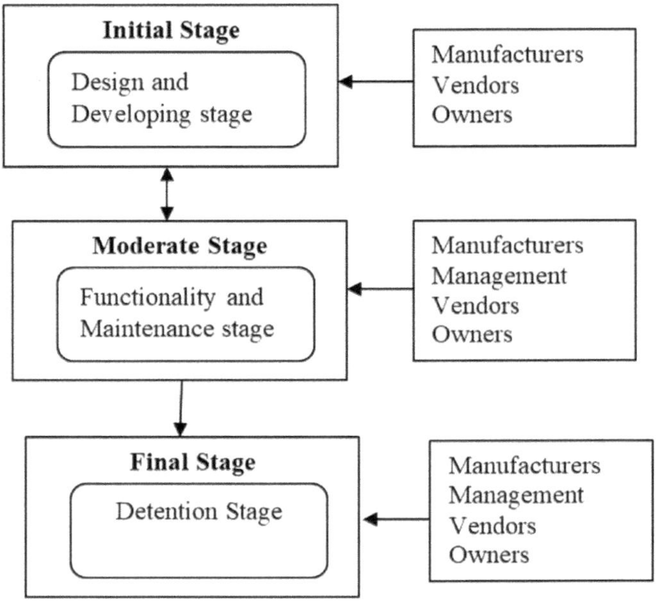

FIGURE 8.3 Lifespan of DTs phases.

8.3.2 CREATING BLOCKCHAIN: SOLUTIONS

In the sections that follow, go over the complex systems that surround smart contracts' innovative concepts (such as choosing the right technology, application specifications, on-off network retention, provenance, and efficient information sources).

1. **Deployment:** Although the majority of current designs merely note a conceptual base built on a blockchain and ignore the basic deployment. Nevertheless, certain projects, like the execution of a model, are only half complete. A fully accessible model that has been extensively developed and is predicated on network methodologies and open-source platforms featuring performance and expense data is currently available. There are few other papers like this in the research, but they don't offer specific technological

information regarding the kind of network and other assessment measures. Likewise, a digital version is impervious to physical fluctuations. However, it is deficient in architectural and other deployment information. It is advised that they create and also develop it. From the other extreme, in contrast to these assessment measures, offering a graphical experience is another crucial part of the application. It is obvious that further attention must be paid to the design in the hopes of recognizing probable endorsement hurdles and opening up new areas for future study because the earlier research have not been fully evaluated in order to acquire further practical insights into the sector. The barrier between the thinker and production might be further closed by the commercial specialists confirming the model's operational applicability.

2. **Categorization:** The architecture and routing algorithm of the network sort may be used to classify them. According to the needs of the set of resources, the network version must be chosen. The majority of the existing DT-based business model uses an open or closed network with a common linear cascaded system as the Virtual Composites framework [15]. This primary cause of tendency indeed is rise in solutions aimed at various application areas, whereas other Blockchain Technologies could not have attained the necessary functional intelligence phase or been increasingly embraced.
3. **Extensibility:** The choice of the suitable blockchain is crucial to evade technical issues that can develop as a result of the upcoming surge based on the increasing number of participants. On-chain methods (raising embedding capacity, better decision techniques, clustering, and network metadata), let alone off solutions, are now available to address the sustainability of distributed ledgers. Sustainability difficulties are tackled by off-chain techniques among the available proof-of-stake DT technologies. The Datagram crypto algorithms, compact string cryptographic protocols, or other designs should always be taken into account while assessing the cryptographic data format.

An autonomously operating code block called a smart contract that operates upon that network can be employed to maintain an unstable party's commitment in order to manage virtual currencies.

In accordance with the demands of the core technology, documents can be used to orchestrate situations. The trades have the potential to manage information exchange procedures to depict the dual activities, to maintain permission data for all people concerned, or to regulate occasion machine connection when implemented in a DT context. Notable instances of consensus mechanism being used in Security mechanisms include calling functions through safety and security rules generally when process conditions fluctuate, adjusting platforms and products settings and monitoring using proactive or reactive self-executing activity management.

4. **Convergence Method:** The stakeholders conduct a consensual process to maintain the blockchain and distributed state synchronization throughout the system by creating a private reliable network among insecure sites [16]. The signature scheme entails tackling an asset, demanding, and indisputably simple issue. As there are other studies discussing various blockchain-based strategies, the behavior of the mechanisms used for resolution is not discussed here. Cryptosystem is implemented in some of the scientific literature on smart contracts algorithms despite offering any compelling justifications.

   Confirmation and Real evidence have both been put to the test to back up the claim that the resolution method should be selected positioned on the fundamental technology. Proof of work has more computational power than Proof of authority. Though, work is more adapted to public blockchain situations whereas work proofing is ideally adapted to conditionally authorized contexts. Use Crossing High Availability, nevertheless, to maximize performance and minimize lag. Likewise, as a component of subsequent modifications to their model, recommended employing without interruption resource-efficient algorithms.

5. **Retention of Data:** Most of the studies conducted on data storage recommend employing offline chain storage, but they do not actually enforce it. It relies on a swarming multicast routing countertop, offline chain storage. By using this strategy, device information might well be retained offline following detector identification online chain while still fulfilling the standards for storage capacity and speed. In contrast to lowering the price of online

chain storage in the conventional network, the concept of offline storage may assist in resolving technical challenges. Nevertheless, there are numerous significant difficulties in merging chain disc space offline with smart contracts related technologies. Off-chain storage necessitates consolidation, realty compatibility including online chain memory, safety concerns, besides no assurance that one off operation is genuine.

6. **Tracking:** Even though the committed nature of smart contracts ensures transparency, choosing the right secret settings is still necessary to monitor and follow the whole chain during the event [17]. Some of the current works describe provenance but don't show how it's used or how it functions in fact. Relying on the starting point and target, the original information relationship, each one vital indicator may exist built during moving along the system and might be saved on the blockchain to rebuild the production system on request. This would impose provenance in the economic structure.
7. **Reputable Sources:** While the network has left behind protective measures, the majority of current activities do not take into account assuring the reliability of information processes and the transmission of messages. Intentionally or accidentally, information counterfeiting can happen for a number of factors, including gadget hacking, information forgery, or intentionally engaging groups, leading to the Waste In, Waste Out phenomenon. Considering the network processes need to ensure the reliability of data at the source, a smart contracts multilayer confidence framework for the IoT was developed to increase edge assurance.

At the protocol stack of the conceptual methodology, the information integrity subsystem captures the level of trust in empirical evidence based on support from nearby different databases, the recognition of information sources based on their behavior in the past, and the degree of trust in the inspection observed by information sources. The necessary actions are recommended to deal with certain problems with smart contract-based electronic problems [18].

- The Internet gadgets need to authenticate, thus the network using their general public keys and cryptographic credentials as approved gadgets in order to prevent gadget interference. By employing these safeguards, the access points can only gather information from approved devices after verifying their legitimacy via the network. Additionally, system upgrades are performed via blockchain technology to prevent viruses or exploit code infiltration.
- To identify the problematic item in the associated activities and to discover irregularities in the platform, which can be detected relying on geographic information systems, are two primary techniques for identifying information falsification during information transfer.
- To recognize a dishonest participating object, it is possible to revoke or withhold membership credentials by performing regular specific place visual examinations of the facilities to ensure that manufacturing activities adhere to moral values and ecologically responsible processes, and ii) implement a financial disciplinary framework to reward honesty while punishing dishonest parties or suspending financial markets' involvement.

8.3.3 IMPLEMENTING SECURITY MECHANISMS

The studies that are currently available on access control architecture concerns involving cryptography, security systems, and atomic resilience are presented in this segment [19].

8.3.3.1 CRYPTOGRAPHY

Many of the literature suggest using encryption to ensure secrecy, but they don't go into much depth or do it right. The lifecycle management information is advised to be protected; nevertheless, the operational specifics of these cryptographic protocols really aren't supplied, which makes it difficult to examine. Identical to how privileges are imposed through securing a key stream with the matching object's encrypted message, cryptography while posting, it is appended to the Cryptographic Map using Advanced Encryption Scheme-256 bit. It is also widely known that machines have the potential to compromise the confidentiality of facto encryption algorithms. Using knock alternatives for general populace security that is nuclear resilient is one approach to solving this important problem. Such solutions, which are focused on fundamental issues apart from doubling and categorical inverse, are nonetheless thought to just be effective from classical assaults. Cryptographic protocols offer yet another variety of options, perhaps significantly Computational Deployment, as a technique for creating keys symmetrically.

8.3.3.2 ACCESS MANAGEMENT

A small amount of permission and verification systems as the corollary of identity management in the existing research on smart contracts-based DTs. The specifics of their deployment, nevertheless, were not available. To allow or restrict entry depending on position on rely on authentication management, a recorded leadership linkage. Together in the comparable pattern, Role-based access control and associated authorization have been merged to form composite network management architecture. It is important to give serious thought to the authorization functionality since it offers identification and permission to permit exchange of information amid the stakeholders.

8.3.3.3 INTRINSIC PERMEABILITY

Only with the advent of computational power, criminals can now fake electronic documents and disrupt operations, endangering the integrity of the blockchain. Smart contract applications that enable computational approaches, like IoTA, should be used to enhance global security. Additionally, blockchain technology needs computational characteristics such as dynamic broadband, computational connectivity, and symmetric key propagation for mutual interaction and exploitation using smart contract networks to improve security and make blockchain faster and more effective [20]. A blockchain alternative known as Twinchain is being studied, replacing Cryptographic Hash Digital Certificates with computationally verified ones. However, the proposed field theory network lacks sufficient information about the selected host security. To address this issue, there are currently open problems and potential research directions that the scientific community could explore.

8.3.4 AUTHENTIC FRAMEWORK FOR BLOCKCHAIN-BASED DTS

It requires an architecture that could demonstrate the reliability of information resources, information during transmission, and stored in order to manage and then extract data from the obtained information [21]. Throughout this context, innovate an authenticity smart contracts DT platform to fulfill the needs of reliable information processing, information protection, and proactive analysis. Figure 8.4 depicts the essential elements of the suggested methodology. Stakeholders in the network, including devices, equipment, and people, initially enroll as authenticated users. Malware localization is then carried out within the protocol stack by tracking, gathering, and analyzing criteria in the external reality that are reflected in the digital realm. When it comes to the reliable functioning of automated machines, the conceptual approach can be quite helpful and is well respected.

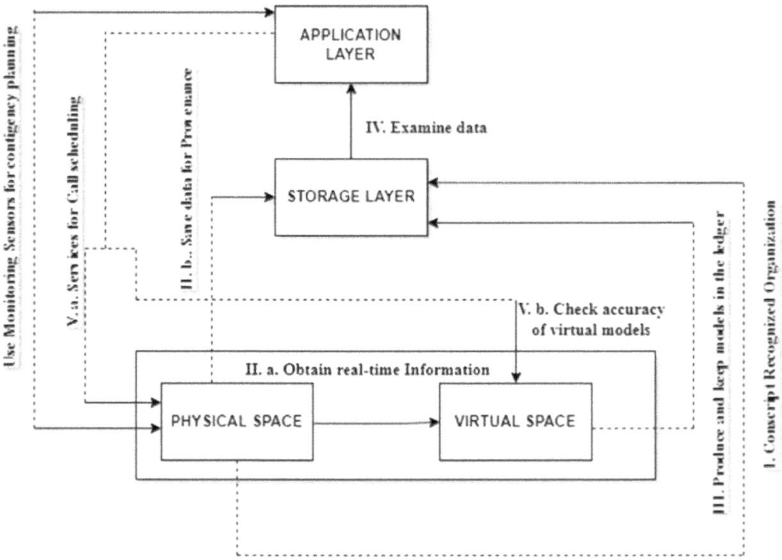

FIGURE 8.4 Authentic framework for blockchain-based DTs.

8.3.5 INTERACTION WITH DTS

Healthcare is a term used recently to describe the deployment of smart manufacturing innovations to medical practices. In order to make medical systems more creative, healthcare solutions in particular strive to build compatible and networked health systems. Huge information is generated by health service operations and is gathered using medical services surveillance instruments such as devices in touch and detectors. Healthcare may be summarized as cooperative clinical computer systems that rely on industrial concepts [22]. As a result, the individual's smart technologies are coordinating and engaging to execute the patient records that are utilized to make informed choices. The condition of the tangible infrastructure is also evaluated via intelligent automation. Therefore, in the application scenario, the stakeholders, which will include judgment, individuals, crew members, and fulfillment, are the hard resources.

8.3.6 PRINCIPAL ADVANTAGES OF BLOCKCHAIN-BASED DTS

While DTs make it possible for industries to realize prospective products and applications in a simple and straightforward manner, these advantages of DTs are predicated on a belief in the dependability, privacy, and authenticity of data. In real-world situations, hackings might happen for a variety of causes, including malevolently and not. Consequently, the information employed by the DTs must be reliable and safe in order to properly exploit the advantages of DT. Utilizing distributed ledger in this way enables businesses to synchronize verified DTs information while simultaneously managing customer databases on a shared database. Therefore, utilizing evaluation factors such as the count of the authors and contributors of the statistics, the existence of a reliable private entity, the recognition of bibliophiles, and the degree of authenticity of the transmitted messages, designers should examine the inquiry of whether blockchain is actually necessary for algorithms. In accordance with the blockchain network model, DT can successfully address these problems because, among other things, the algorithmic components of the blockchain absolutely promise information robustness and the blockchain's key agreement offers responsibility and visibility. This section goes over the main advantages of employing blockchain technology which are described in subsections.

8.3.6.1 AUTHORITATIVE SOURCES

Utilizing IoT solutions that are potentially susceptible to cyber-attacks, shaping industries that relate the material world to the virtual world is made possible [23]. Because the gathered information drives sensible policy, it's crucial to demonstrate the reliability of the methods used to obtain the information. By gathering information about enrolled gadgets to proactively mitigate gadget manipulation or by leveraging a smart contract tiered security framework for estimating telemetry information, technology could potentially play a vital role.

8.3.6.2 DECENTRALIZED PLATFORM

The existence of DT is divided into stages during which several participants perform possible stances upon this. Such situations are not particularly adapted for the designs of data storage processing. Furthermore, that is too complicated to organize the electronic pair process because of the independent information silos, particularly while in instance of digitized strands is necessary to connect data across different life cycle stages. Blockchain technology may be used to provide large-scale data storage without a centralized control point. As a result, the blockchain network may be used to archive and retrieve other materialistic of information and also DT evidence such as traceability information moving throughout various layers.

8.3.6.3 DATA PROTECTION FOR DTS

While DTs must be based on reliable data, rising automation and connectedness create new strategic directions that might endanger business resources or procedures. It is necessary to find a solution for the issue of how to guarantee the reliability and authenticity of information gathered from the external reality and then shared with the virtual environment. Blockchain's immutable and irrevocable properties have the potential to reliably collect data from several collaborating entities. The network with technological capability preserves the irreversible records of transactions involving the production or accessibility to DTs. Additionally, the combination of openness and confidentiality may be facilitated through the utilization of aliases during network processes, like authentication tokens for expressing network personalities. To protect their anonymity, permit secure information exchange, and combat covert operations seeking comparative strategy, networks might protect several monikers. As a result, several unauthorized ecosystem participants can exchange DTs information by maintaining its security, consistency, and reliability.

8.3.6.4 INFORMATION TRACING

Furthermore, authenticity facilitates smart contracts via DTs may infer information about the locations when a given dataset at any given time in the production process [24]. Furthermore, beyond information analysis, it is possible to consult the whole history of DTs, from its inception to the moment, in order to determine what generates differences among the virtual realms.

8.3.6.5 ACCESS RIGHTS

Throughout the instance of both the corporate sector, entitlements apply to both viewing and editing documents. Similar reasoning is valid for DT in addition to the permitted parties who are allowed to manage data, which can only be accessed by a chosen group of individuals. Smart contracts authorization manages the DT-specific regulations and guidelines. Additionally, blockchain automatically stores the background of alterations, which can avoid the issue of unauthorized information manipulation, causing additional issues with information protection.

8.3.6.6 DATA PROTECTION

The association of various DTs linked to various highly substituted should be reinforced by twins in order to enable an analytical perspective of the whole events. A central authority called a Detachable Twin Implementation replicates a tangible object as well as function, as opposed to a simulation model aggregation is the combining of twin instance, ideally comparable instance, to replicate the material reality as a whole universe [25]. The DT Aggregation is a valuable option for assaults and might have serious overall effects. Goods may be exposed to information leaks. By allowing, smart contract DTs enable sufficient protection mechanisms to safeguard DT instances within the aggregation. In addition to maintaining information trustworthiness, solely authorized personnel may modify.

8.3.6.7 AUTOMATION

Mechanization is necessary for production activities to start scheduled and flawless activities. Recording information on actual phase transition and their equivalents is based on the fundamental specifications and software applications as from the network can be utilized to perform situations.

8.3.6.8 TAKING ON THE ISSUE OF COUNTERFEITS

Producing knockoff goods that are almost exact reproductions of the genuine goods has a negative effect on the identity and credibility of the items together with the investor expenditures in the sector. By supplying the necessary integrity, the technology and DT in conjunction can guarantee that scammers are recognized and prevented. Most specifically, the openness whereas the digitizing of the data comes from the network. Capital and goods will be backed by the smart contracts. As a result, by eradicating the issue of knockoffs, the combination of DT and blockchain technology will be crucial in ensuring the financial return for both the enterprises. Furthermore, smart contracts that rely on DT solutions can offer a genuine authentication process to address significant issues with the validity of DTs by detecting forgers who seek to copy or distribute inauthentic DTs.

8.4 INCORPORATING AI AND IOT WITH BLOCKCHAIN-BASED DTS

Though DTs provide many benefits, they might also have drawbacks. The use of actual information analysis helps to derive meaningful insights from the vast amount of information. Despite all the explanatory accuracy underlying deep learning techniques, AI may help overcome these constraints. As shown in Figure 8.5, blockchain, along with AI and IoT, can deliver trustworthy and efficient DTs. However, there are several factors to consider when incorporating intelligence with smart contracts [26]. This subsection offers a pathway to intelligence, blockchain-based DTs. The rationale for AI, along with IoT in DTs and blockchain, is where humans come in. It then provides a comprehensive look at the role of intelligence with IoT in DTs based on blockchain and the unresolved issues that need to be addressed.

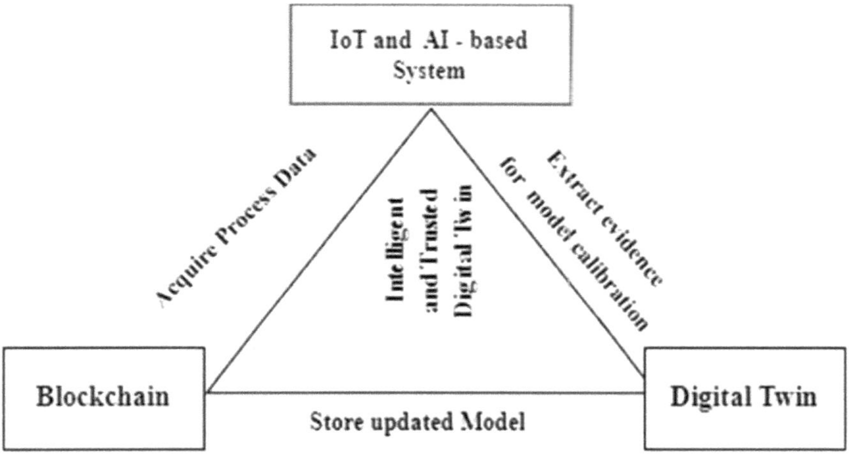

FIGURE 8.5 Incorporating blockchain along with IoT and AI for DTs.

8.4.1 DT SOLUTIONS-BASED ON INTELLIGENT DEVICES

Due to the incorporation of several unreliable yet interrelated parties involved with fragmented information from various origins, the dynamic, varied, and linked structure of sophisticated complex processes makes it vulnerable to mistakes and assaults. Bidirectional or consecutive interconnection of operations [27], forecasting the impact of seemingly harmless acts or occurrences may be harder as a result of the fundamental system's growing interconnection. Neglecting to take into account these modest changes in models' function might lead to the program's implosion or have a significant impact on its protracted activity.

Incorporating intelligent technologies in this aspect can assist in reducing cyber threats by exploiting potential risks, flawed concepts, malicious attacks, current attack pathways, and automatic surveillance systems to recognize negligible assaults. DTs are used to generate test information. However, business-specific structural elements are used to create annotated outliers. To put it more simply, an inadequate classification algorithm uses fictitious information to identify irregularities in the actual facts. It is wise to use Classifiers in smart contracts along with synchronous replication and fine-tuning capabilities to guarantee that their detached modules can distinguish whether they are being fed fraudulent or inaccurate extrinsic information, notice the onset of such an assault, and then continue to a protected mission in the presence of the assault.

To guarantee the accuracy of the protection regulations, the account should always be recorded and accessible. A smart contracts system with authenticity understanding may monitor and identify the responsible party for implementing or amending the safety rules. In accordance with the educate and protect method, a ledger-based DTs controlled by an intelligence automation platform could trigger an alert to the necessary protective systems when it recognizes the existence of hostile or unintentional interruptions inside the infrastructure. The intelligence approach, although the injury is inevitable, minimally separates the destroyed actual (micro) element, protecting the remainder of both structures from additional destructive assaults.

8.4.2 QUALITY OF INFORMATION ENHANCEMENT

Besides making predictions, intelligence contributes to enhancing the consistency of the information by using sophisticated machine learning algorithms to populate incomplete information to smooth out the results. The Information Grand Cherokee approach, particularly using intelligence information filtering, could assist in obtaining information correctness and completion, which eventually enhances machine learning and efficiency through the linked impact. Therefore, discrepancies might be substituted using deep convolutional systems to enhance the accuracy of the information. Comparable to extracted features, the proposed method uses little additions to actual information to increase the quantity of useful information [28]. Information recorded from several varied sources may be combined using techniques like neural net models, Gaussian filters, and probabilistic reasoning to deliver reliable and thorough interpretations of the event.

8.4.3 METRICS IN DTS WITH INTELLIGENCE ASSISTANCE

Computation, forecasts, or risk assessment of a network using DTs may be done utilizing classifiers, unstructured, and other learning techniques. Based on the amount of both classified and unclassified knowledge in DTs, the built-to-take approach may be adopted. Nevertheless, machine learning might well be applied in these information circumstances. Podcasts may be made on the blockchain to identify patterns in specific forecasting; one needs to log the results and capture the learning parameters and evaluation metrics.

8.4.4 BLOCKCHAIN-BASED DTS: INTELLIGENCE CHALLENGES

Regardless of the advantages of using AI that have been discussed, there are nevertheless issues that must be resolved until intelligence of smart contracts relied on DTs are materialized. In general, researchers concentrate primarily on those information obstacles that intelligence in smart contracts [29]. To circumvent intelligence errors in these contexts, it is crucial to recognize these issues.

1. **Imperative of Explicit Intelligence:** The majority of classifiers are a contraption; therefore, the comprehensive accomplishment of intelligence technology that is capable of seeing, understanding, anticipating, and operating independently still remains an unwillingness to employ those systems in fact. Actions produced by the associated technologies pose a greater risk and narrative coherence, especially to individuals who aren't intelligence specialists, which might have a detrimental effect on the platform's credibility. Regarding protection platforms, where complexity is vital since even a tiny malfunction in an automated method within decision-making processes might result in incorrect conclusions, uncertainty is much higher.

   Within processes and enhancements, which can signify a tragedy or a financial meltdown. Moreover, personal information empowerment advocates autonomous intelligence despite ensuring the validity, reliability, and accountability of both the information and computations. Throughout this aspect, combining intelligence with blockchain promotes control systems in intelligence via monitoring the flow of information and automation devices have complicated tendencies that can improve comprehension and trust in judgments. Through the underpinning mechanisms, one such use of automation is collaborative classification techniques. It calls for the information provenance distributed ledgers promoted in order to hide cyber threats.

2. **Information Constraint:** The essence of machine learning algorithms is information. A substantial volume of information is essential in order to understand the characteristics and generate correct forecasting, and this need frequently turns into an obstacle in intelligence. From an extreme sense, there seem to be different options for processing information: gathering data from actual external conditions and creating information artificially. Information silos created by digital companies or unwillingness to exchange data due to rivalry might be other causes of high dimensionality. Information shortage should be considered when using intelligence within smart contracts since DTs use archival and actual information to recreate or duplicate the external world.

   To supply the AI systems, homogeneous, constant, wholesome, and efficient information streams need to be provided. Leveraging people to publicize knowledge using smart contracts information exchange through a digital sector is such a potential remedy for the issue of high dimensionality. Throughout the initial stages (construction and implementation) of DT existence, interested parties could employ the generated information. Related to this, aggregated information might offer fictitious parameters for compatibility mapping, especially if the system's present crystalline structure is undetermined. By mentoring new versions just on participants' gadgets using the recipients' own information, distributed computing may be employed to tackle the difficulties associated with evidence bottleneck.

3. **Prototype Bias and Erroneous Issues:** The discrimination issue and prototype failures are inextricably linked to the information used to train intelligence. Cognitive development might be skewed and lead to forecasting mistakes if the intelligence algorithm is developed on excessive or substandard information. This reduces

the effectiveness of machine learning. Although information gathered from external reality is utilized for learning, it is crucial to consider the information's origin, surroundings, interpretation, and method of gathering it in order to prevent bias in the results and subsequent prototype mistakes.

By motivating users to submit high-quality information and collecting evidence from many fragmented providers using smart contracts, participants learned that the use of technology can tackle the skew in facts and prototype fault difficulties [30]. It is significant to highlight that the network architecture used and the inherent motivation for providing information will determine the allocative efficiency of doing so. Exchanging information from various sources alone does not quite address the issue of statistical imbalance. To address the issue of informational skew, ensemble learning techniques with integrity must be added to hashing algorithms. Additionally, consensus protocols and autonomous platforms may be used to engage collaborative judgment mechanisms. By revisiting the decision-making process and providing the necessary arguments, authorized personnel may reevaluate their faith in a process and determine if the problem lies with the system or the user.

8.5 CHALLENGES IN IMPLEMENTATION AND INVESTIGATION

The underlying scientific, operational, and ethical issues of smart contracts rely on DTs strategies needing more research in the drive to impose reliable information handling. The unresolved issues that prevent the broad integration using smart contracts assist DTs in the sector are identified. Several difficulties bring up unique opportunities for this line of scientific enhancement. The comprehensive categorization of the existing and upcoming study and application problems for smart contracts-based DTs.

8.5.1 SCIENTIFIC CHALLENGES

The technological difficulties related to the synthesis of DTs and cryptographic protocols are as follows:

1. **Constraints on Information:** Among the main obstacles to creating a digital simulation that accurately depicts the realistic thing is creating an elevated model. Adverse situations occur as a result of the fundamental semi component, vitality, and diversity, faulty devices, user errors, covert flaws, data corruption, research shortcomings, and the unpredictability, fuzziness, and instability of tangible infrastructure. Additionally, if the actual item changes with time, a matching development of the simulations is necessary while keeping interoperability. Consequently, a practical approach is required by including intellectual DT infusion in order to help system uncover the possible outcomes.

   Abstractions or modifications of characteristics are conceivable when taking into account generalizing and elaboration in classifiers, provided that they don't alter the performance in the particular circumstance and inside the given setting. Restoration mechanisms are employed in this context to monitor material things and replicate actions by imitating their response. In beginning the regeneration activity in the infrastructure or perhaps the client limitations with geography exist. All versions of a tangible product must be in the exact state as the primary, according to the period specified by the program. If not, it can cause an issue with knowledge coordination and the integrity among the real item and its several duplicates.

   Leveraging smart contracts to combine activity business services becomes a potential explanation to these problems [31]. Contextual constraints and the requirement for source information in the prior to create the very same triggers are two additional problems that duplication may encounter. Additionally, complicated and interconnected challenges need to be solved by professionals. In order to underline the consequences of social economic development and to close misunderstandings between subject authorities, Sentient Job Information Architecture and End Framework collaboration uniting technical people, developers, and customer interaction shortages. Seamlessly in this sense, techniques must be outfitted with very well and easily used web applications that blend in with the capital product and enable the terminal to produce, alter, verify, and manage multimedia content in such an acceptable manner.

The possible four reasons are the major causes of information difficulties: The set of information to obtain, pace of information gathering, data processing, and integration of various data types [32]. The primary problem necessitates an exchange of large quantities of information and little variety of information, as both can lead to erroneous projections if there is insufficient analysis, driven decision stagnation if there is too much data. The rate at which the specimen gathering is the problem. Lesser errors might be detected whenever generator oscillations were recorded, but testing a single minute may provide excess information and slow down the translation. Additionally, the length of time for which information is held must be decided, as lengthy information is essential for analyzing yet big numbers of quick assessments result in significant operational expenses. Consequently, it is crucial to specify the geospatial precision of the information in accordance with the infrastructure needs.

The difficulty has to do with enormous information preservation on the network, which necessitates deep review of things like the significant expense of maintaining large amounts of quick assessments, memory requirement, and speed. Progression capacity, maintaining merely information codes to cut inventory level plus detachable network architectures are some external prevention techniques and strategies. Considering those fixes, there are currently unresolved concerns with the huge information processing difficulty due to the amount and transfer speed [33].

Sound waves, degrees, recordings, and graphics all have distinct kinds, protocols, architectures, and dimensions, making it difficult to combine them. In such cases, durations might not be in synchronization throughout data screening at various rates, combining delicate information might result in the deletion of features, and misplaced or inaccurate data can skew conclusions and mask flaws. In order to integrate practical and systematic data into the environment, additional research is needed to address informational difficulties such as quantity, diversity, truthfulness, production flow, spectral granularity for sensor information, and appropriate handling, preservation, and recovery.

2. **Deficit Spending:** Incorporating the latest innovative developments like the Internet, supercomputers, hyper ledger, virtual twins, artificial intelligence, and others helps businesses cut costs, enhance production, offer manufacturing equipment with forecast and current obligations, and improve serviceable productivity. Nevertheless, as a result of these developments, businesses must invest a significant volume of funds in the ongoing administration of interconnected equipment. The economic feasibility must be calibrated for these kinds of procedures. Alternatively, it can deplete the technological infrastructure required to handle corporate operations and business investments. Many businesses are aware of the importance of blockchain, artificial intelligence, and IoT in lowering threats, strengthening protection, and optimizing processes, but they also understand that perhaps the internet, computation, throughput, and capacity needed to analyze enormous amounts of information associated with DT development are outrageously expensive.

   Organizations may face major obstacles as a result of blockchain technology, anticipated dangers related to the advancements, huge stage of beginning deployment expenses, and the potential for upsetting operational practices [34]. The crucial certainty decisions and operations might otherwise be based on false findings due to incorrect or absent impacting circumstances. In certain cases, intermediate- to huge enterprises stand to gain more from the adoption of the fresh framework than micro-enterprises. Nonetheless, they might probably grow or as a result of current technology, micro-businesses must embrace them as quickly as possible to advance technology.

3. **Interoperability Approaches and Difficulties:** A genuinely worldwide architecture and foundation for DTs are still far off owing to a scarcity of regulatory requirements, the unavailability of consistent data and prototype standardization, and compatibility issues in corporate activities. Mismatches often remain during the whole natural process of growth. Distinct information gateways, unique operating systems, class specifications, multiplexing, and collaborative learning are all involved in the dispersed control of information throughout the life cycle of a project. Additionally, it

consists of several suppliers, freelancers, and clients of varying sizes and professions that may not always use the same operating system. Accordingly, the need for better interconnection at each stage of the lifecycle ensures that safety assets from the preceding phase are preserved and generates reusable traces, which is the driving force behind the creation of the DT standard based on the consecutive stages.

A general trademark creation and progress infrastructure for DTs with a convenient application programming interface should be created, having clear rules for artistic disciplines with interoperability, enabling these designs to function without the need for big data management and committed contractor alliances. Blockchain networks have one thing in common, which depends on a global, autonomous people-to-people meshwork. The fundamental file format, expansion approaches, high availability, extent of dispersion, and decision methods differ greatly among them. Knowledge constraints extremely dispersed economic environments, complete in it and unconnected networks, to name a few drawbacks of non-concord deploying blockchain.

4. **Controversy About Distributed Ledger Technology:** Problems regarding the pragmatic use of the distributed ledger within the sector are being expressed by a plethora of recent studies conducted. A few of one's anxieties will be how to guarantee the consistency of information from parties involved with sensing devices given the correlation among the virtual and electronic realms, delay problems only with a growing quantity of base stations, insufficient knowledge resulting in pale industries, surveillance troubles, inefficient security concern, and the garbage in, garbage out problem, judgment dysfunction associated with data overstimulation, and energy requirements.

   While supporting the complete spectrum of commercial operations, distributed ledger technology struggles to solve issues. However, issues resulting from the hidden nature of distributed ledgers encourage individuals to place their faith in the legitimacy and authenticity of the procedures without becoming aware of the underlying technology. For this purpose, businesses are wary of implementing distributed ledger in classifiers. Distributed ledgers may be acquired, and their applications examined by individuals through lectures, seminars, manuals, and credentials offered by research institutions and maker spaces.

8.6 CONCLUSION

The development of DT replicas would have enabled analysis, forecasting, and enhancement of all processes prior to their execution within the physical realm. Decentralized information architecture is required due to data being gathered and retrieved through disparate origins of information and structures held by many selected firms. Utilizing smart contracts in this way enables businesses to handle data on a public blockchain that safely logs product portfolio data processing activities. As a consequence, DTs enhance serviceability by prioritizing problem mitigation over alternative approaches, whereas blockchain technology facilitates the development of reliable DT architecture.

In this investigation, a thorough analysis of the planning and deployment problems with the most recent, cutting-edge smart contracts relies on DTs. As a result, the range of applications and advantages of DTs has continuously expanded. Smart hackers, meanwhile, have the capability to breach connected devices and cause irreversible harm to the connected applications and also to the associated infrastructure. Comparable to how intelligent device communication may be compromised to acquire private information while it is in movement or at rest.

KEYWORDS

- **artificial intelligence**
- **blockchain**
- **cryptography**
- **data protection**
- **digital twins**
- **internet of things**
- **lifecycle**

REFERENCES

1. Saeed, H., Alex, V., & Svetlana, V. (2023). Blockchain-empowered security and energy efficiency of drone swarm consensus for environment exploration. *IEEE Transactions on Green Communications and Networking, 7*, 328–338.
2. Thang, P., Trang, T., & Tai, B. (2022). Multimedia privacy, security, and protection within the blockchain: A review. In *Proceedings of the 5th IEEE International Conference on Contemporary Computing and Informatics* (pp. 1171–1176).
3. Kavitha, M., & Roobini, S. (2023). Systematic view and impact of artificial intelligence in smart healthcare systems: Principles, challenges, and applications. In *Machine Learning and Artificial Intelligence in Healthcare Systems* (pp. 25–56).
4. Yasin, C., Ioan, P., & Yacine, R. (2021). Leveraging BIM and blockchain for digital twins. In *IEEE International Conference on Engineering, Technology and Innovation* (pp. 21–28).
5. Sandner, P., Jonas, G., & Robert, R. (2020). Convergence of blockchain, IoT, and AI. *Frontiers in Blockchain, 3*, 1045–1055.
6. Bhaskar, K. B., Prasanth, A., & Saranya, P. (2022). An energy-efficient blockchain approach for secure communication in IoT-enabled electric vehicles. *International Journal of Communication Systems, 35*, 1–20.
7. Radhya, S., Saeed, H., Kenneth, N., & Bader, A. (2022). Blockchain-based digital twins' collaboration for smart pandemic alerting: Decentralized COVID-19 pandemic alerting use case. *Hindawi Computational Intelligence and Neuroscience, 6*, 234–244.
8. Volodymyr, B., & Gilles, H. (2020). OM Forum–Distributed ledgers and operations: What operations management researchers should know about blockchain technology. *Manufacturing & Service Operations Management, 22*, 223–428.
9. Barbara, R., Elena, C., & Daniela, F. (2019). A survey on digital twin: Definitions, characteristics, applications, and design implications. *IEEE Access, 7*, 167653–167671.
10. Volkan, D., Raja, J., Guntur, D., & Ali, D. (2019). A trust architecture for blockchain in IoT. In *EAI International Conference on Mobile and Ubiquitous Systems: Computing, Networking and Services* (pp. 190–199).
11. Marietheres, D., & Günther, P. (2020). Unleashing the digital twin's potential for ICS security. *IEEE Security & Privacy, 18*, 20–27.
12. Marietheres, D., Benedikt, P., & Günther, P. (2019). A distributed ledger approach to digital twin secure data sharing. In *IFIP Annual Conference on Data and Applications Security and Privacy* (Vol. 11559, pp. 281–300).
13. Dinh, T., & Thai, M. (2018). AI and blockchain: A disruptive integration. *IEEE Computer Society, 9*, 48–53.
14. Dorri, A., & Jurdak, R. (2020). Tree-chain: A fast lightweight consensus algorithm for IoT applications. *IEEE Computer Society, 15*, 369–372.
15. Ali, D., Salil, S., Raja, J., & Praveen, G. (2019). LSB: A lightweight scalable blockchain for IoT security and anonymity. *Journal of Parallel and Distributed Computing, 134*, 180–197.
16. Gao, R., Wang, L., & Teti, R. (2015). Cloud-enabled prognosis for manufacturing. *CIRP Annals, 64*, 749–772.

17. Haya, R., Khaled, S., Raja, J., & Mohammed, O. (2020). A blockchain-based approach for the creation of digital twins. *IEEE Access, 8*, 230–238.
18. Henry, R., Herzberg, A., & Kate, A. (2018). Blockchain access privacy: Challenges and directions. *IEEE Security & Privacy, 16*, 38–45.
19. Sihan, H., Guoxin, W., Yan, Y., & Xiongbing, F. (2020). Blockchain-based data management for digital twin of product. *Journal of Manufacturing Systems, 54*, 361–371.
20. Hussain, F., Hussain, R., & Hossain, E. (2021). Explainable artificial intelligence (XAI): An engineering perspective. *IEEE Access, 9*, 394–411.
21. Muhammad, I., Pankesh, P., John, G., & Ramy, H. (2021). Cognitive digital twins for smart manufacturing. *IEEE Intelligent Systems, 36*, 96–100.
22. Justin, D., & Harris, B. (2019). Decentralized & collaborative AI on blockchain. In *IEEE International Conference on Blockchain* (pp. 14–17).
23. Khan, A., Shahid, F., Maple, C., & Jeon, G. (2020). Towards smart manufacturing using spiral digital twin framework and Twin chain. *IEEE Transactions on Industrial Informatics, 1*, 99–107.
24. Andrew, K. (2017). Smart manufacturing must embrace big data. *Springer Nature, 544*, 23–25.
25. Jiewu, L., Douxi, Y., Qiang, L., & Kailin, X. (2019). ManuChain: Combining permissioned blockchain with a holistic optimization model as bi-level intelligence for smart manufacturing. *IEEE Transactions on Systems, Man, and Cybernetics: Systems, 50*, 182–192.
26. Claudio, M., Antonio, M., Gianluca, P., & Andrea, U. (2019). Building a digital twin for additive manufacturing through the exploitation of blockchain: A case analysis of the aircraft industry. *Computers in Industry, 7*, 134–152.
27. Miller, D. (2018). Blockchain and the Internet of Things in the industrial sector. *IT Professional, 3*, 15–18.
28. Minerva, R., & Crespi, N. (2021). Digital twins: Properties, software frameworks, and application scenarios. *IT Professional, 23*, 51–55.
29. Benedikt, P., Marietheres, D., Philip, E., & Günther, P. (2021). EtherTwin: Blockchain-based secure digital twin information management. *Information Processing & Management, 1*, 58–80.
30. Huang, S., Wang, G., Yan, Y., & Fang, X. (2020). Blockchain-based data management for digital twins of product. *Journal of Manufacturing Systems, 54*, 361–371.
31. Lischke, M., & Fabian, B. (2016). Analyzing the Bitcoin network: The first four years. *Future Internet, 8*, 567–575.
32. Viriyasitavat, W., & Pungpapong, V. (2019). Blockchain and Internet of Things for modern business process in digital economy—The state of the art. *IEEE Transactions on Computing Social Systems, 6*, 1420–1432.
33. Feng, Y., Zhou, M., Tian, G., & Li, Z. (2019). Target disassembly sequencing and scheme evaluation for CNC machine tools using improved multi-objective ant colony algorithm and fuzzy integral. *IEEE Transactions on System Management Cyber Systems, 49*, 2438–2451.
34. Karve, P., Guo, Y., Kapusuzoglu, B., & Mahadevan, S. (2020). Digital twin approach for damage-tolerant mission planning under uncertainty. *Engineering Fracture Mechanics, 3*, 225–235.

CHAPTER 9

Collaboration of Blockchain-Based Digital Twins in the Healthcare Industry: From Idea to Implementation

G. BOOPATHI RAJA,[1] M. PARIMALA DEVI,[1] T. SATHYA,[1] and ARUN SEERALAN BALAKRISHNAN[2]

[1]*Department of Electronics and Communication Engineering, Velalar College of Engineering and Technology, Tamil Nadu, India*

[2]*Department of Mechatronics Engineering, Asia Pacific University of Technology and Innovation, Kuala Lumpur, Malaysia*

ABSTRACT

Digital twins (DTs) have been made possible by the quick development of computer, storage, communications, and networking technology. It is undoubtedly the digitized version of any physical feature, item, or machine. Further, it can be used for prototyping, simulation, validation, and 3-D design before the real component is built. A DT can be employed for configuration, surveillance, diagnostics, and prognostics once a physical component has been put into service. It is anticipated that the Internet of Things (IoT) and DTs will attract a lot of attention soon and will be crucial for the Healthcare Industry. However, current DT construction techniques, platforms, and tools are heavily centralized and unable to provide trustworthy data provenance, audit, and traceability. The information about transactions, logs, and history is neither safe nor impenetrable. A blockchain-based development process for DTs has been designed to enable secure and reliable traceability, availability, and data integrity of

Blockchain-Based Digital Twins: Research Trends and Challenges.
A. Prasanth, Vandana Sharma, Rajesh Kumar Dhanaraj, Pethuru Raj (Eds.)

transactions, records, and data provenance. Few methods employ smart contracts to control and monitor actions taken by those participating in DT generation. This chapter deals with the analysis of Blockchain-based DT for healthcare applications. It also provides the various integration constraints and challenges faced in the healthcare Industry.

9.1 INTRODUCTION

An online representation of a real-world organization, item, or service is known as a digital twin. A wind farm, a jet engine, as well as a larger entity like a skyscraper or even a whole city, might have a digital twin. The physical item is modeled as a digital twin in this digital era. To collect data or information and predict the operation of physical assets and process the same, digital twin technology may be employed. It is also used to recreate both of these representations [1].

The strength of digital twins derives from the ability to better view real-world assets and data by linking them together [2]. Cross-functional teams can work together on the interactive and immersive design, development, testing, deployment, and operation of complex systems because of digital twins. They support businesses in understanding the history, seeing the current, and averting issues in the future. Insights from sales and marketing, assessment, 3D visualization, simulation, and forecasting are used to inform decision-making [3]. A digital twin is often computer software that uses data from the real world to replicate how a product or service might function. These programs can boost output using industry 4.0 technologies such as software analytics, artificial intelligence, and the IoT [4–6].

Digital twins are created by integrating conceptual models using Building Information Modeling (BIM) and Geographic Information System (GIS) or scanning them to examine and analyze actual items in the real world with business and IoT data. Various data sources, including information and models, may be gathered, organized, and presented as realistic, interactive visualizations using a real-time 3D-enabled digital twin. A computer graphics technology called real-time 3D produces dynamic content faster than the human eye can see it.

Digital twins are representations in the virtual realm of the loads, connections, and movements that assets may encounter in the physical world. Customers may engage with dynamic, three-dimensional, and responsive content in real-time. They may instantly view the findings on any platform, including portable devices, computers, and augmented, mixed, and virtual reality (AR/MR/VR) devices and they can precisely simulate real-world events, what-if scenarios, and any situation conceivable in the virtual environment. Every deployment of a digital twin is different. The complexity and business effect of deployments often increase with each stage that is added. A DT can be anything from a 3D product configuration to an accurate depiction of a city-sized network or system with each of its parts dynamically connected to engineering, construction, and operating data.

The virtual models have emerged as a foundation in contemporary engineering to promote innovation and boost efficiency as a result of the development of machine learning as well as other factors including big data. In other words, having one might help see important technological trends, and prevent expensive physical item breakdowns and testing procedures and services utilizing improved quantitative, monitoring, and forecasting skills.

Virtual representations of patients (referred to as "health digital twins") are created from the database of different categories of patients, their statistical medical data, and real-time modifications of patient and environmental factors are known as "physical twins" in the medical field. Human Digital Twin (HDT) can be used to simulate random impacts on the digital twin to offer insight into the predicted performance of the physical twin, presenting a ground-breaking potential in personalized medicine, diagnostic tests, and healthcare. Computational needs, data governance, clinical implementation, and product supervision are important things that need to be taken care of when converting HDT studies into clinical practice [7].

9.2 DEFINITION OF A DIGITAL TWIN

The digital twin is a process of modeling a system or a service that extends its lifecycle, which incorporates modeling, machine learning, and inference to assist decision-making and is maintained with real-time data [8, 9]. It is divided into three parts:

- an object in a real-time environment;
- the digital twin is expressed in a software form; and
- data that connects the first two components.

A virtual model serves as an intermediary representation of the actual state of the thing it represents. Furthermore, it is specific to the item it represents rather than being merely general. The digital twins of two things that appear to be similar are frequently not the same [10, 11]. Figure 9.1 shows the process for defining a Digital Twin. Although a 2D or 3D CAD image is frequently connected with a digital twin, it is not necessary. A database, a list of equations, or a spreadsheet might make up the digital representation or digital model.

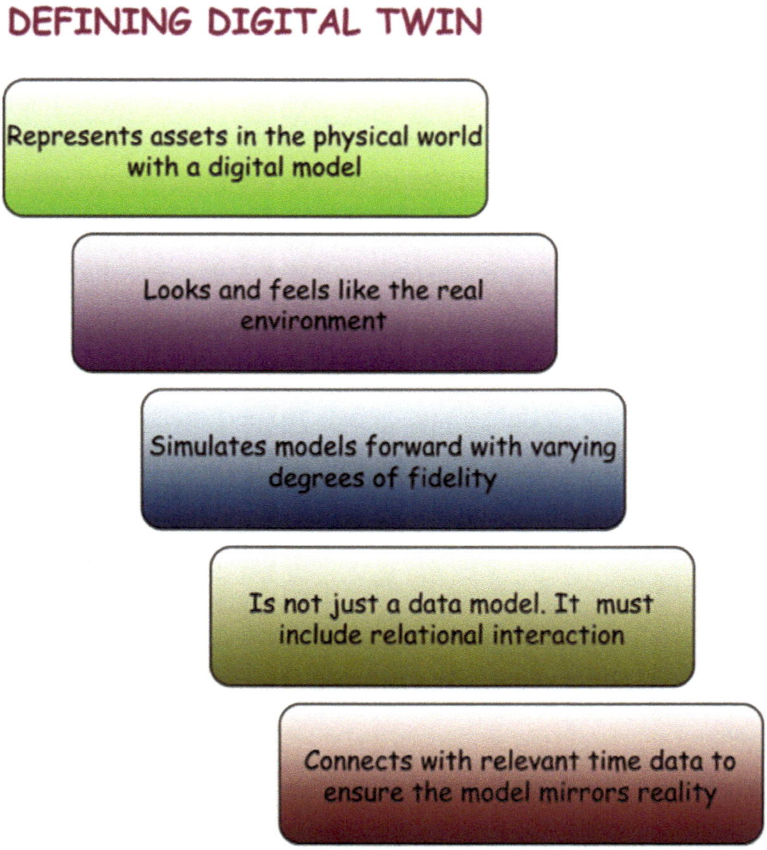

FIGURE 9.1 Process for defining digital twin.

Digital twins are distinct from related notions due to the data link, which is frequently but not always two-way. Users may query the data to learn more about the condition of the item or process through this link, and actions expressed through the digital twin can affect its physical counterpart [12]. The key addition to the original definition is synchronized at a set frequency and quality according to the Digital Twin Consortium, an organization that promotes standards and works to develop the market.

Three important characteristics of the technology are mentioned in these qualifications:

- The goal of synchronization is to ensure that the represented thing and its digital duplicate are as similar as feasible.
- Depending on the goal, a digital twin's update frequency or pace can be quite variable, ranging from seconds to weeks to on-demand.
- Fidelity is the level of accuracy and precision of the synchronization mechanism and the virtual representation.

Designers and engineers utilize the same modeling and CAD tools to create digital twins throughout the initial product development process. The distinction between the digital twin and the traditional twin is that the digital twin is handled via Product Lifecycle Management software and retains and updates the model for use in subsequent phases of the product's lifespan such as inspection and maintenance.

IoT sensors are frequently used as the physical links between the thing and its twin, and a developed IoT implementation is frequently needed for digital twins [13]. Digital twins are claimed to be equally important to IoT since they give otherwise chaotic and challenging-to-interpret IoT data structure, analytics, and usability.

9.2.1 PROCESS OF DIGITAL TWIN

A virtual picture created to perfectly replicate a real-world object is called a digital twin. The object under evaluation, such as a wind turbine, is equipped with a variety of sensors that have been interconnected to crucial functioning areas. These sensors generate information on the characteristics of the physical device, including temperature, power capacity, and other parameters as well as environmental factors. This knowledge is subsequently applied to the digital copy by the processing system. With this information, the system prototype may be used to do simulations, look into performance concerns, and propose changes, each to produce informative data that can subsequently be transmitted to the actual device [14]. Figure 9.2 shows the model for representing Digital Twin.

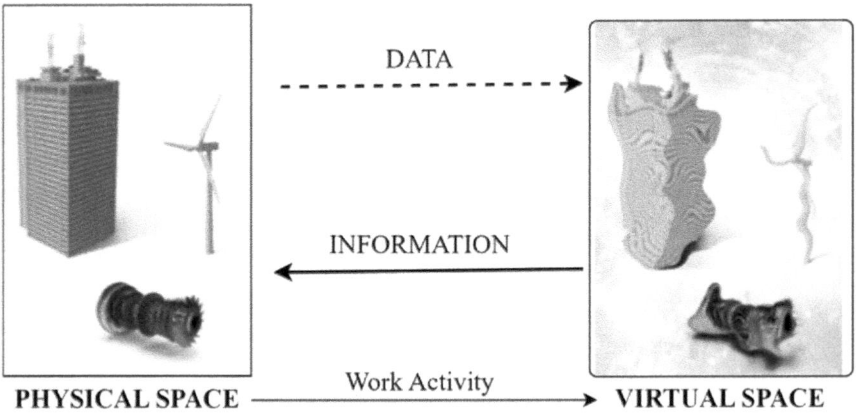

FIGURE 9.2 Digital twin model.

9.2.2 *SIMULATION VERSUS DIGITAL TWIN*

Digital twins and simulations may utilize the virtual model to replicate the multiple operations of any system. The digital twin may be more helpful for study because it is essentially a virtual scenario. A simulation and a digital twin differ mostly in scale. A digital twin may contain as many relevant simulations as necessary to simultaneously investigate several processes, whereas simulations normally only analyze one procedure at a time. There are likely more differences. For example, real-time data is frequently not advantageous for simulations.

Digital twins, on the other hand, are created to work with multiple information flows that start whenever the system processor receives crucial data from object sensors. This continues whenever the processor exchanges further information with the source object [15]. By providing more accurate and up-to-date data from various disciplines, digital twins enhance the computational capability offered by a virtual environment [16]. Since digital twins can examine more issues from a wider variety of angles than normal simulations can, they have a higher potential to ultimately improve products and processes [17].

9.3 HISTORY OF DIGITAL TWINS

The digital twin technology concept was first introduced in 1991 by David Gelernter in the book "Mirror Worlds." Dr. Michael Grieves, a professor at the University of Michigan, developed the idea of DT software and implemented it in 2002. John Vickers of NASA then coined the term "digital twin" in 2010.

However, the basic concept of studying a physical thing with a digital doppelganger was developed much earlier. Figure 9.3 shows the roadmap for the advancement of technology in the electronic field.

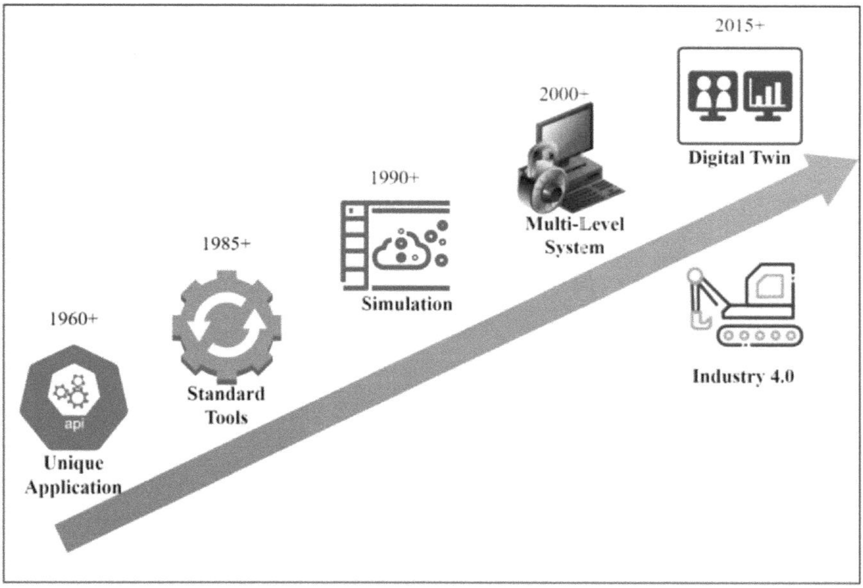

FIGURE 9.3 Roadmap of the advancement of technology.

9.3.1 ORIGIN OF THE IDEA OF DIGITAL TWINS AND IMPLEMENTATION

After Challenge Advisory sponsored a talk on technology at the University of Michigan in 2002, the concept of a digital twin became well-known. The creation of a product lifecycle management center was the subject of the presentation. It included all of the digital twin's components, including actual and virtual spaces as well as the diffusion of information and data flow between them. The idea of combining a physical and digital twin into a single entity hasn't changed over time, even though language may evolve. Digital twin technology was developed in the 1960s, contrary to popular belief, which holds that it was first utilized in 2002. Figure 9.4 shows various stages of Digital Twin Implementation.

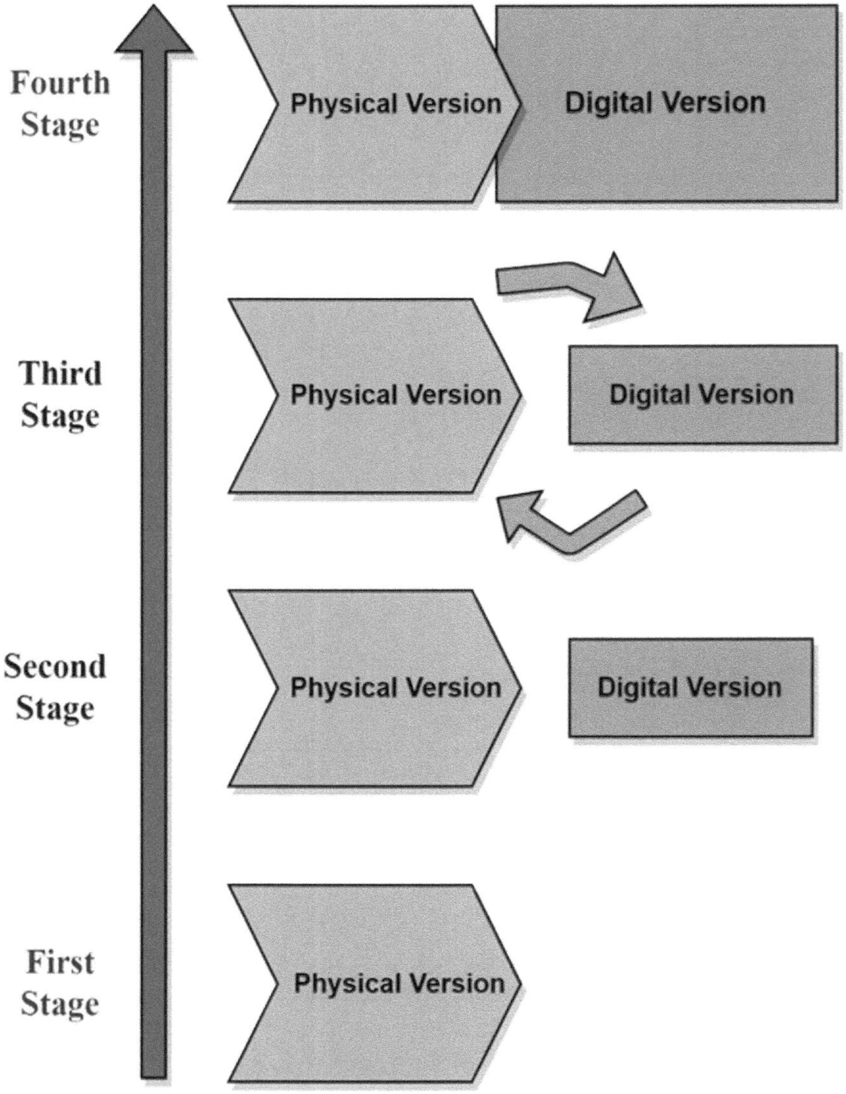

FIGURE 9.4 Stages of digital twin implementation.

NASA would use fundamental twinning principles during this period to program spacecraft. Researchers built systems that have been physically equivalent to those in space to replicate the systems in space. Nobody could have predicted that after the oxygen tanks exploded, the Apollo 13 mission, which had been launched in April 1970, might swiftly turn into a battle for survival. Technical problems had to be rectified from up to 2,00,000 miles away, making it a legendary rescue mission as everyone held their breath. The fact that NASA possessed a digital twin replica of Apollo 13 on Earth allowed engineers to test potential fixes from ground level, which was a crucial aspect of the rescue effort. Systems are now mostly virtual simulations rather than actual ones, of course.

9.4 TYPES OF DIGITAL TWINS

Digital twins may be many different things, based on how much the technology has been improved. The primary area of operation where these twins diverge has emerged. It is typical for many kinds of digital twins to work together in a system or process:

1. **Component Twins/Parts Twins:** A component twin, which is the most compact representation of an operational component, is the essential building block of a digital twin. Twin components are almost identical yet relate to far less important elements.
2. **Process Twins:** The macro-level of magnification used by the process twins shows how several systems work together to build a full manufacturing complex. Using process twins, it may be possible to identify the precise temporal patterns that eventually affect overall efficacy.
3. **Asset Twins:** Any time two or more things function successfully together, an asset is produced. Asset twins allow researchers to study the interactions between these factors. It produces a large amount of operational data that may be analyzed to yield valuable insights.
4. **System or Unit Twins:** The next phase in the magnification process is to construct system or unit twins. It helps consumers understand how various parts interact to form an entire working

system. System twins reveal how assets interact and may point to performance enhancements.

Three major categories of digital twins can be distinguished, each of which illustrates a different use for the technique:

i. **Digital Twin Prototype (DTP):** Before a tangible item is created, something is done;
ii. **Digital Twin Instance (DTI):** This is done following the creation of a product to test it in various use scenarios.
iii. **Digital Twin Aggregate (DTA):** This program compiles DTI data to assess a product's capabilities, perform prognostics, and evaluate operational conditions.

The broad categories may be used to accomplish a variety of activities, including system planning, product creation and redesign, logistical planning, and quality control and management. A digital twin can be used to reduce time and costs whenever testing is necessary for a product or process, whether for design, deployment, monitoring, or improvement.

9.4.1 LEVELS OF DIGITAL TWINS

Five different stages of complexity exist for digital twins. The most complex models are capable of functioning independently, while the simplest models just incorporate input from numerous sources.

- **Stage 1:** Descriptive Twin: It is a representation of a manufactured thing and is a dynamic, modifiable form of construction and design data. The type of information that users want to add and the type of data that they want to remove are specified by the users.
- **Stage 2:** Informative Twin: Functional and sensory information is available as an additional layer at this level. For guaranteed system communication, the twin gathers, aggregates, and validates the required data.
- **Stage 3:** Predictive Twin: Insights may be gained by this twin using operational data. For example, consider how an automobile might alert customers when it needs a change of oil.

- **Stage 4:** Comprehensive Twin: It analyzes hypothetical situations and models possible futures.
- **Stage 5:** Autonomous Twin: On behalf of users, this twin can assemble data and take appropriate action. It is crucial to emphasize that Levels 1 and 2 were still in operation for Architecture, Engineering, and Construction (AEC). The IoT and in-built sensors have improved Levels 3, 4, and 5 by adding real-time data.

9.4.2 ADVANTAGES OF DIGITAL TWINS

The benefits of DTs include both long and short-term. Digital twins will allow BIM to continue to provide value throughout the asset's lifespan in the long run. The focus of immediate gains is on operations, offering a single point of reference that could assist in saving costs, lowering risk and uncertainty, and boosting productivity.

1. **For Owners:** Smarter Decisions, Better Return on Investment: Owners have a difficult time managing, monitoring, and ultimately maximizing the value of their facilities when data is scattered among separate systems. Users may experience scenarios and get extensive data because of digital twins. It will maximize performance and lengthen the lifespan of their assets, assisting them in making smarter business decisions from the planning and design phases on.
2. **Smoother Operations and Simpler Maintenance for Managers:** Without having to fight with ambiguous (or nonexistent) maintenance documentation, owners and operators may begin running assets effectively right away. Owners and operators have an unsolvable problem while developing systems integrated with all forms for monitoring, controlling, and optimizing their benefits due to the transmission of analog, unclassified, and disconnected data. Due to this, owners and operators of smart buildings are unable to benefit from them and are instead left with disjointed data, erroneous information, and limited knowledge of how their assets are working. To manage things including maintenance and optimizing energy use, DTs offer a digital thread which integrates cumulative data in a straightforward manner, enabling operators to concentrate on their responsibilities.

Data from facilities are always changing. Managers can see who is making adjustments and thanks to digital twins. To save waste and improve safety, managers can even forecast material and labor cycles.

3. **More Value Corresponds to More Business for AEC Firms:** AEC companies that can provide more services and value win more contracts. Robert Bray, the general manager of Autodesk Tandem, claims that the company's AEC clientele has the potential to provide owners with additional information. Additionally, if the asset is digital and depicts the built representation of the model, it may link to operating systems and assist individuals with issues following handover.

   Operations yield close to 80% of an asset's lifetime value. AEC companies add more value by helping customers get the most out of their investment through increasing energy savings, improving maintenance, and enhancing occupant quality of life when they can work with owners to establish performance priorities and expectations upfront.

4. **For All:** The Long-Term Value: Designers and owners may deliberate for a relatively long time about the functioning of their assets and broader consequences as digital twins gain traction in AEC. Future cities may be built on digital twins to be more resistant to threats such as pandemics that might strike at any time, the planet's rapid population increases, and diminishing natural resources.

9.4.3 EXAMPLES OF DIGITAL TWIN TECHNOLOGY

Businesses utilize digital twin technology for a wide range of purposes, including streamlining existing operations, training staff, and testing new goods or procedures before actually launching them in the real world, where they are more expensive and difficult to rectify any flaws.

- To investigate the operations model portrayed by the digital twin, machine learning and artificial intelligence techniques are routinely used, regardless of wherever the actual facility is placed, even if the hardware is in space. To operate, maintain, and repair equipment while you are not physically close to them, NASA has employed pairing technology, the forerunner to digital twin technology, since the beginning of space travel. The Apollo 13 mission was saved in precisely this fashion by engineers and astronauts on Earth. Now, NASA uses digital twins to investigate future vehicles and planes.

- They have the chance to mimic the application before it goes live; digital twin technology can assist staff in becoming familiar with the Internet of Things installation and automation. The physical environment can then benefit from the insights gained and possibilities discovered through a digital twin.
- Providing a real-time image of what is occurring with machines or other physical assets, digital twins have proven to be very helpful in manufacturing to decrease maintenance concerns and assure optimal production productivity. By using digital twin technology on machinery in oil fields and refineries, Chevron anticipates saving millions of dollars in repairs and maintenance by 2024.
- Digital twins might revolutionize patient care and healthcare processes. They can help in manufacturing. Surgery and other medical operations can be carried out on mannequins instead of real people due to the usage of digital twins of patients or organs. Patients may be monitored by tiny sensors the size of bandages, and these digital models can be used for treatment planning and AI monitoring.
- The four human hosts of the Spring Festival Gala, which was broadcast on China Central Television, were joined by an AI version of themselves that ObEN had constructed. They are more than computer-generated avatars as they use natural language processing, machine learning, and computer vision to construct virtual representations of the hosts. The development of AI-powered educators, healthcare workers, and nurses is one of the technology's potential future applications.
- Even the racing of Formula 1 cars has benefited from the application of digital twin technology. A simulation can assist the driver and the vehicle crew in determining what tweaks can enhance performance in a sport where each second counts.
- Even Singapore has a digital doppelganger. Consider all the factors that affect municipal management. City planners may better understand and manage energy use with the aid of digital twin technology, which also has a variety of other uses that can enhance the quality of life for city residents.

9.4.4 FEATURES OF DIGITAL TWINS

1. **Greater Efficiency:** Before a new product enters production, digital twins can help in monitoring or mirroring production processes to attain and maintain maximum efficiency during the entire manufacturing process.
2. **Better R&D:** Due to the volume of data generated on expected performance results, the usage of DTs facilitates efficient product research and design. Before beginning production, organizations may be able to make necessary product improvements with the use of this data and insights.
3. **Product End-of-Life:** Digital twins can be used by producers to inform decisions on how to handle items that are now approaching the end of their useful lives and require final processing, such as recycling or other processes. They can decide which production materials to buy by employing digital twins.
4. **Businesses and Enterprises Using Digital Twins:** The adoption of digital twins is not required for all products or businesses, although they have many benefits. Not every item has the level of complexity that calls for the continual, intense inflow of sensor data that digital twins require. Furthermore, it may not always be advantageous to spend a significant sum of money on the development of a digital twin.
5. **Better Availability and Dependability:** As a practical method to boost system performance, monitoring, simulating, and controlling a network or asset is a good idea.
6. **Reduced Risk:** By decreasing asset and process-related events and preventing unscheduled downtime, researchers may safeguard the environment, the health and safety of the employees, and the company's goals.
7. **Reduced Maintenance Expenses:** Identify problems before they become breakdowns, place component orders, and plan repairs at periods that won't interfere with production objectives.
8. **Enhanced Production:** Ensure product quality by having real-time visibility into the operation of assets and processes to drive customization, respond to it, and reduce supply chain effects.

9. **Faster Time to Value:** The utilization of the most comprehensive Blueprint collection, easy-to-use tools, and industry knowledge was considered to provide the analytics and real-time capabilities that the industry requires.

   However, there are certainly significant benefits to adopting digital models for various applications:

   - Initiatives requiring a sizable physical footprint, including the construction of buildings, bridges, or other complex structures must adhere to strict technical standards.
   - Complex mechanical activities such as those involving vehicles, aircraft, and jet engines. Digital twins can help make large engines and complex machines more productive.
   - Power equipment – This category covers both the systems for producing and transferring power.
   - Projects involving manufacturing – Digital twins are fantastic at increasing process efficiency, which can be seen in industrial settings using cooperating machine systems.

   The following industries that deal with big items or projects are consequently where digital twins are most successful:

- Engineering (systems);
- Manufacturing sector;
- Power utilities;
- Railcar design;
- Automobile manufacturing;
- Aircraft production;
- Building construction and civil works.

9.5 RESEARCH CHALLENGES

Large-scale data collection is one thing, but intelligent data consumption is quite another. Data is used to make the best decisions, but the quality of the data depends on the capacity to use it to model and forecast business situations. Every business undergoing a digital transformation runs the danger of becoming overrun with raw data before discovering a means to analyze and utilize it. Capturing raw data these days is less difficult than processing it, removing the superfluous components, integrating it, and turning it into information that the user can understand inside the framework of their application [18, 19]. The primary obstacle is releasing the power of knowledge. Enterprise and IoT data has already been submerged in spreadsheets, databases, and models [20–22].

9.5.1 DIGITAL TWIN MARKET AND INDUSTRIES

1. **Using Digital Twins to Increase Productivity:** End-to-end digital twins enable the entrepreneur to increase output while decreasing equipment downtime.
2. **The Market for Digital Twins:** Poised for Development: Although they are now employed in a wide range of sectors, the market for digital twins is expanding swiftly, suggesting that there will be a continued need for them [23]. During the year 2020, the digital twin's market share was expected to be 3.1 billion USD. Some industry experts predict it will expand quickly through at least 2026, reaching a predicted value of USD 48.2 billion [24, 25].

9.5.2 APPLICATIONS

Digital twins are broadly used in the following applications:

1. **Equipment for Generating Power:** Large-sized engines like jet engines, power-generation turbines, and locomotive engines benefit greatly from the usage of digital twins, particularly when deciding if periodic maintenance is necessary.
2. **Systematic and Structures:** Large physical structures including imposing skyscrapers or offshore drilling rigs might benefit from using digital twins, especially during the design phase. The HVAC systems that are already employed in those structures can be designed with their assistance.
3. **Manufacturing Operations:** Digital twins have flourished at all phases of production, from the initialization phase to the final product stage and all operations in between phases, since they are created to imitate a product's whole lifespan.

4. **Healthcare Services:** Digital twins may be used to profile patients seeking medical care, exactly as they have been with products. The tracking of several health markers and the generation of important insights may be done using the same sort of sensor-generated data system.
5. **Automotive Industry:** Digital twins are frequently employed to boost vehicle efficiency in the automotive division and reduce production costs. Vehicles are a collection of complexes, interrelated systems.
6. **Urban Planning:** Digital twins might show spatial data in 3D and 4D in real-time. It incorporates augmented reality innovations into constructed environments. This is very beneficial to structural engineers and other experts working on urban planning projects.

9.6 IMPACT OF DIGITAL TWIN ON THE HEALTHCARE INDUSTRY

Over the next three years, digital twin investments are expected to rise, according to 66% of healthcare executives. This is because digital twins boost an organization's performance in the healthcare sector, enable identifying areas for improvement, allow for customized and individualized treatment and diagnostics, and facilitate the creation of new drugs and medical equipment. Figure 9.5 shows the functions of Digital Twin technology in the Healthcare area.

9.6.1 HOW CAN DIGITAL TWIN TECHNOLOGY APPLY TO THE MEDICAL FIELD?

A digital twin is an exact reproduction of the equipment, personnel, procedures, and systems used by enterprises. Using computer models, digital twins are used in healthcare systems to create digital representations of healthcare data, including hospital surroundings, lab findings, human physiology, etc. Virtual twins are created using information about the person, population features, and environment.

9.6.2 ROLE OF THE DIGITAL TWIN IN HOSPITAL

Digital twin technology is used to build a virtual twin of hospitals. Operational strategies, allocation of resources, staffing, and healthcare service models may be studied to identify areas for improvement, foresee future challenges, and fortify organizational plans. As a result, hospital digital twins may be utilized to create facility copies, which in turn allows for:

1. **Resource Optimization:** Hospital administration is better able to spot bed shortages, organize staff schedules, and manage spaces by integrating history and current data on hospital operations as well as the surrounding region (such as car accidents, COVID-19 cases, etc.) to develop digital twins. By improving resource efficiency and optimizing hospital performance for both patients and staff, expenses may be reduced. According to review research, for instance, a hospital was able to shorten the time it took to treat stroke patients by using digital twins to manage the efficient coordination of multiple procedures.

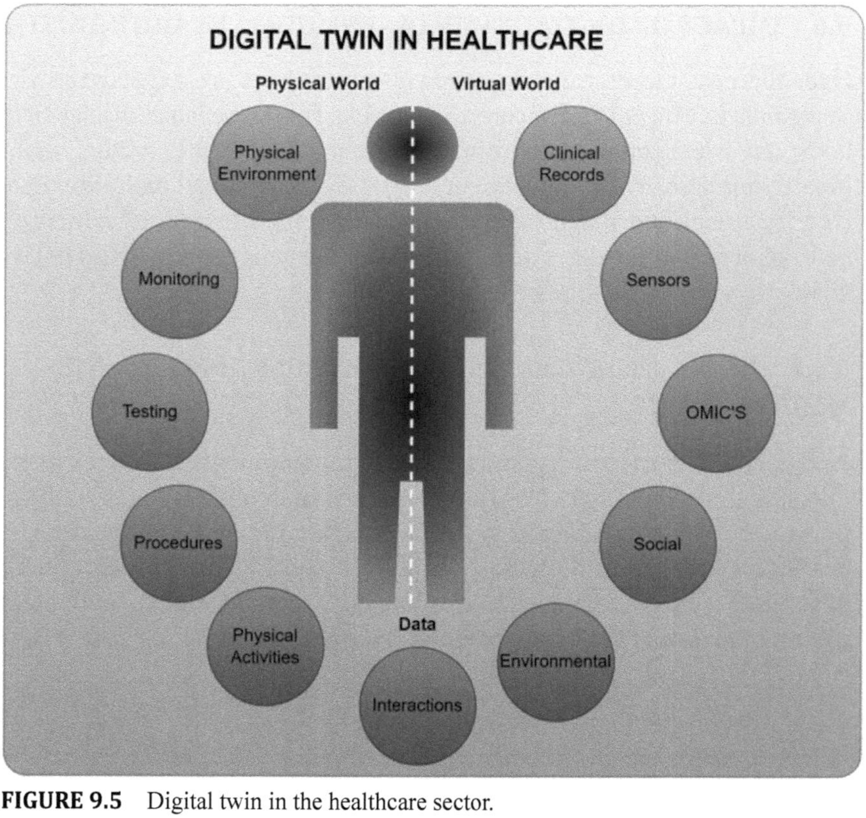

FIGURE 9.5 Digital twin in the healthcare sector.

2. **Risk Management:** To make data-driven strategy choices in a delicate and challenging context, digital twins provide a safe environment to assess changes in system performance (staffing levels, operating theater vacancies, device maintenance, etc.).

9.6.3 DIGITAL TWIN OF THE HUMAN BODY

Digital twins are also employed to imitate organs and single cells in addition to a person's genetic makeup, physiological features, and lifestyle decisions to produce tailored drug and treatment plans. By using such models of the internal organ systems in the human body, medical treatment, and patient care are enhanced.

1. **Personalized Diagnosis:** Vital information (such as oxygen levels, blood pressure, etc.) may be collected and used at the individual level by using digital twins. This device helps users track persistent diseases and, as a result, overall priorities and contacts with doctors by supplying basic knowledge. Thus, clinical trials and laboratory research data are built on such personalized data.

   To choose the optimal course of therapy, therapists avoid utilizing large samples by concentrating on each patient individually. Instead, they track how each patient responds to various therapies using individualized simulations, which increases the accuracy of the whole treatment plan. Despite the interest in and growing volume of attempts for personalized treatment, there are presently no uses for digital twins for actual patients.

2. **Treatment Planning:** Using sophisticated models of the human body, clinicians may find pathology before diseases become noticeable, explore remedies, and enhance surgical readiness.

9.6.3.1 EMPLOYING DIGITAL TWINS TO DESIGN NEW DRUGS AND DEVICES

The creation, testing, and oversight of novel pharmaceuticals and medical equipment can be enhanced with the use of digital twins. For instance:

1. **Medications:** Using digital twins of chemical compounds and pharmaceuticals, researchers may redesign or change drugs while

taking into account particle size and composition properties to enhance delivery effectiveness.

2. **Medical Devices:** Developers can examine a product's functions or features, alter its composition or design, and then evaluate the effects of changes in a virtual setting before the product is produced. This reduces the costs related to failures while significantly enhancing the performance as well as the safety of the end product.

9.6.4 DIGITAL TWIN CHALLENGES IN HEALTHCARE

In the healthcare sector, the application of digital twin technology now faces many challenges, including the ones listed below:

1. **Limited Adoption:** Digital twin technologies have not yet been fully incorporated into clinical practice routines. The use of technology in healthcare facilities (such as hospitals and labs) should be increased to improve clinical procedures, digital simulations, and patient care in general. However, despite the growing usage of digital twins in the healthcare system, it is suggested that they will continue to be pricey and out of reach for most people. Digital twin technology will start to be seen as a perk available only to those with more financial means, leading to inequalities in the healthcare system.
2. **Data Quality:** Artificially intelligent digital twins can adapt from the available biological data, but they are constantly aware that the quality of the data may be low because it was gathered by for-profit organizations. As a result, it is difficult to analyze and portray such data. In the end, this has a severe impact on the models, which also has a bad impact on the model's dependability throughout the diagnostic and treatment procedures.
3. **Data Privacy:** As a result of the deployment of digital twins, data on each individual will need to be gathered more often by healthcare and insurance companies. These health organizations progressively get a full awareness of an individual's biological, genetic, physiological, and lifestyle-related information. Such

unique data may be exploited for commercial objectives as opposed to people's interests. One example would be an insurance company using the data to identify individuals with greater accuracy.

9.6.5 PRECISION MEDICINE TECHNIQUES USING HEALTH DIGITAL TWINS

Digital twins can now forecast events beyond what is possible with traditional predictor technologies, such as ICU IoT-enabled full physiological monitoring, owing to the convergence of AI, IoT, and CLO. This HDT architecture represents a basic shift from established large medical data analytical methods like logistic correlations and regressions. HDT systems create virtual twins using deep phenotyping, or the automated analysis and fusion of distributed data gathered from biological data, patient records, and portable sensors. The patient reaction to the medicine, behavior modification, and environmental variables may be measured and then anticipated using the physical twin data.

Prediction analysis for real-time applications introduces additional chances for accurate prevention and treatment systems, such as the identification and severity grading of abdominal aortic aneurysms. Figure 9.6 shows the various Decision support schemes preferred at the time of Patient Diagnosis and Treatment.

Data fusion of enormous data from various sources, including contradictory information, is essential for the creation of AI-driven customized treatment. Effective data classification and exchange certainly face difficulties from a number of these sources, including Electronic Health Record data and imaging reports. To overcome these integration challenges, researchers in HDT have to apply predefined pre-processing data algorithms or techniques including signals and natural language processing, etc. Data fusion methods are also capable of accounting for the rising processing complexity of real-time client records, including those generated by wearable technology. For instance, the process can be sped up by performing the least complex data integration on the wearables. Wearable sensors are used to export-related data to the centralized information system. When developing data management methods, HDT technology should ideally consider these factors.

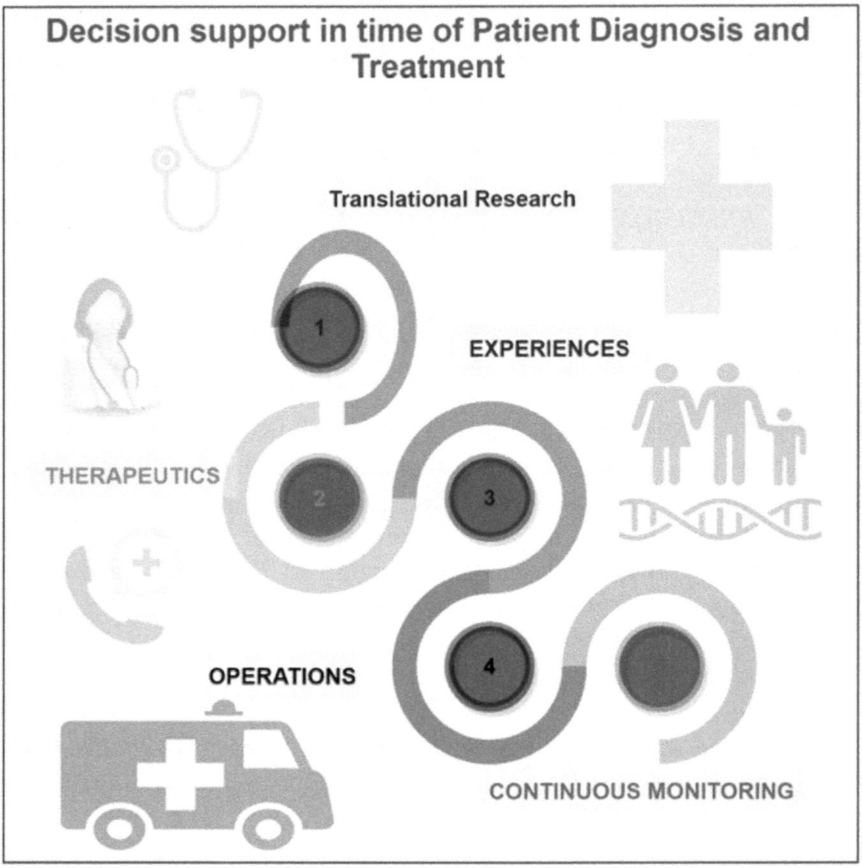

FIGURE 9.6 Decision support at the time of patient diagnosis and treatment.

Payment plays a significant part in the implementation of these systems. There are numerous ways to finance HDT advancements. First, if providers efficiently utilize HDTs, they could just be a necessary evil. The application of the HDT system supports patient monitoring and treatment. It may lead to larger margins for doctors who participate in payment models based on value, such as periodic and huge payments. Patients will face an increase in repayment which occurs due to the adoption of HDTs. If the same is adopted, patient-centered and process-related services are acquired by them in a quick span. Insurance companies have to think about looking outside the conventional fee-for-service paradigm to find payment schemes. It effectively encourages the use of elevated HDTs without leading to overuse that raises healthcare costs, due to the almost least marginal expense which utilizes Artificial Intelligence-based systems including HDTs.

9.7 FUTURE OF DIGITAL TWIN

There is a significant shift occurring in current operating paradigms. Industries that strongly rely on physical assets are going through a disruptive digital transformation that necessitates a combined physical and digital perspective of all the tools, facilities, and operational processes that are accessible. Digital twins are a key element in such reconfiguration. Because more and more cognitive resources are being dedicated to their utilization, digital twins have nearly unlimited potential. Along with continuing to produce the insights required to improve goods and streamline operations, digital twins must continually develop new abilities.

Products, machinery, industries, structures, and communities are no longer only physical objects; they now have precise virtual analogs. Digital twins of persons exist as well. In the metaverse, users will encounter the next generation of the internet in real-time 3D, along with the interconnectedness of systems, gadgets, and people. The metaverse is creating an entirely new market having endless possibilities for interactive content in cross-digital and blended reality domains, whether they are used to manage facility renovations or customize a vehicle purchase. One such economy is 3D marketing.

Researchers are pleased to introduce people to their summit on the future of digital twins, which is taking place as the top trend in digital twins continues in 2018. System components, programming models, or digital twins are expected to represent billions of things within a few years. By 2020, it is anticipated that over 21 billion digitally connected sensors and 50% of big industrial enterprises will be using digital twins, which may result in billions of dollars in maintenance, repair, and operating savings. Additionally, it is anticipated that up to 60% of manufacturers will observe the performance of the system through the DTs and the quality of products by the year 2020.

Digital twins will be utilized by up to 60% of international businesses to improve customer service. The proliferation of IoT-designed sensors is driving the employability of digital twins and the future of both is intertwined. The operation and reaction of an object to its environment may be revealed by sensors and the usage of digital twins can improve analysis, condition modeling, operations, and value. Digital twins are important, and this fact suggests that they will continue to expand and be used for billions of years to come.

9.8 CONCLUSION

The combination of real-time data collection and AI-powered analysis found in HDT technology moves healthcare one step closer to completely tailored therapy. The larger group of patients present in HDTs is utilized to simplify clinical trials and guide the users to select a proper health policy in addition to individual patients. Data fusion accuracy, patient and provider literacy, remuneration, and governance of HDTs and associated secondary data economy will be crucial factors to take into account as HDTs begin to permeate clinical workflows. Integration of digital twin technology, blockchain, and AI technologies deliver smart personalized healthcare services to all human beings. Thus, digital twin technology was an essential tool in revolutionizing the healthcare Industry and its standards.

KEYWORDS

- **artificial intelligence**
- **blockchain**
- **data fusion accuracy**
- **digital twin**
- **healthcare industry**
- **internet of things**
- **real-time data collection**

REFERENCES

1. Rasheed, A., San, O., & Kvamsdal, T. (2020). Digital twin: Values, challenges, and enablers from a modeling perspective. *IEEE Access, 8*, 21980–22012.
2. Venkatesh, K. P., Raza, M. M., & Kvedar, J. C. (2022). Health digital twins as tools for precision medicine: Considerations for computation, implementation, and regulation. *NPJ Digital Medicine, 5*, 150.
3. Muthusamy, P. D., Velusamy, G., Thandavan, S., & Govindasamy, B. R. (2022). Industrial Internet of things-based solar photovoltaic cell waste management in next-generation industries. *Environmental Science and Pollution Research, 29*(24), 35542–35556.
4. Parimala Devi, M., Raja, G. B., Gowrishankar, V., & Sathya, T. (2020). IoMT-Based Smart Diagnostic/Therapeutic Kit for Pandemic Patients. In C. Chakraborty (Ed.), *Internet of Medical Things for Smart Healthcare: Covid-19 Pandemic* (pp. 141–165).
5. Parimala Devi, M., Choudhry, M. D., Nithiavathy, R., & Boopathi Raja, G. (2022). Blockchain-Based Edge Information Systems Frameworks for Industrial IoT: A Novel Approach. In S. Misra (Ed.), *Blockchain Applications in the Smart Era* (pp. 19–39). Springer International Publishing.
6. Kavitha, M., & Roobini, S. (2023). Systematic View and Impact of Artificial Intelligence in Smart Healthcare Systems, Principles, Challenges, and Applications. In *Machine Learning and Artificial Intelligence in Healthcare Systems* (pp. 25–56).
7. Coorey, G., Figtree, G. A., Fletcher, D. F., & Snelson, V. J. (2022). The health digital twin to tackle cardiovascular disease—A review of an emerging interdisciplinary field. *NPJ Digital Medicine, 5*(1), 126.
8. Barricelli, B. R., Casiraghi, E., & Fogli, D. (2019). A Survey on digital twin: Definitions, characteristics, applications, and design implications. *IEEE Access, 7*, 167653–167671.
9. Chakshu, N. K., Sazonov, I., & Nithiarasu, P. (2021). Towards enabling a cardiovascular digital twin for human systemic circulation using inverse analysis. *Biomechanics and Modeling in Mechanobiology, 20*(2), 449–465.
10. Attaran, M., & Celik, B. G. (2023). Digital Twin: Benefits, use cases, challenges, and opportunities. *Decision Analytics Journal, 6*, 100165.
11. Azzaoui, A. E., Kim, T. W., Loia, V., & Park, J. H. (2021). Blockchain-based secure digital twin framework for a smart healthy city. In *Advanced Multimedia and Ubiquitous Engineering* (pp. 107–113).
12. Dilsizian, M. E., & Siegel, E. L. (2018). Machine meets biology: A primer on artificial intelligence in cardiology and cardiac imaging. *Current Cardiology Reports, 20*, 1–7.
13. Jayachitra, S., Prasanth, A., & Hariprasath, S. (2023). AI-Enabled Internet of Medical Things in Smart Healthcare. In *AI Models for Blockchain-Based Intelligent Networks in IoT Systems: Concepts, Methodologies, Tools, and Applications* (pp. 141–161).
14. Singh, R. P., Hom, G. L., Abramoff, M. D., & Campbell, J. P. (2020). Current challenges and barriers to real-world artificial intelligence adoption for the healthcare system, provider, and the patient. *Translational Vision Science & Technology, 9*, 45.
15. Parikh, R. B., & Helmchen, L. A. (2022). Paying for artificial intelligence in medicine. *NPJ Digital Medicine, 5*, 63.

16. Kaplan, A. V., Baim, D. S., Smith, J. J., & Feigal, D. A. (2004). Medical device development: From prototype to regulatory approval. *Circulation, 109*(25), 3068–3072.
17. Kamel Boulos, M. N., & Zhang, P. (2021). Digital twins: From personalized medicine to precision public health. *Journal of Personalized Medicine, 11*, 745.
18. Lo, C. K., Chen, C. H., & Ray, Y. Z. (2021). A review of Digital Twin in product design and development. *Advanced Engineering Informatics, 48*, 1–10.
19. Liu, M., Fang, S., Dong, H., & Xu, C. (2021). Review of Digital Twin about concepts, technologies, and industrial applications. *Journal of Manufacturing Systems, 58*, 346–361.
20. Jia, P., Wang, X., & Shen, X. (2023). Accurate and efficient digital twin construction using concurrent end-to-end synchronization and multi-attribute data resampling. *IEEE Internet of Things Journal, 10*, 4857–4870.
21. Sekar, J., Aruchamy, P., & Abdul, H. (2022). An efficient clinical support system for heart disease prediction using TANFIS classifier. *Computational Intelligence, 38*, 610–640.
22. Shohin, A., Xun, X., Ray, Z., & Yuqian, L. U. (2021). Digital Twin as a Service (DTaaS) in Industry 4.0: An Architecture Reference Model. *Advanced Engineering Informatics, 47*, 1–15.
23. Leng, J., Wang, D., Shen, W., Li, X., & Liu, Q. (2021). Digital twins-based smart manufacturing system design in Industry 4.0: A review. *Journal of Manufacturing Systems, 60*, 119–137.
24. Alazab, M., Khan, L. U., Koppu, S., & Ramu, S. P. (2022). Digital twins for healthcare 4.0- recent advances, architecture, and open challenges. *IEEE Consumer Electronics Magazine*, 1–8.
25. Chen, Z., & Huang, L. (2021). Digital twins for information-sharing in remanufacturing supply chain: A review. *Energy, 220*, 119712.

CHAPTER 10

Blockchain-Based Digital Twin for Predictive Maintenance of Machines Using Machine Learning

D. VINODHA,[1] E. A. MARY ANITA,[1]
PADMAVATHY THIRUNAVUKKARASU,[2] and P. GEETHA[2]

[1]CHRIST (Deemed to be University), Bangalore, Karnataka, India

[2]Department of Computational Intelligence, SRM Institute of Science and Technology, Kattakulathur, Tamil Nadu, India

ABSTRACT

Digital Twin (DT) is a software replica of the physical asset or process that serves the automobile industry by empowering engineers and data scientists to gain a deeper insight into the system and solve the biggest challenges with less time and cost. Blockchain technology is an evolving and promising technology that foresees providing surveillance facilities for DTs, resulting in firming up security and improving the openness, reliability, and unchangeability of DTs. By incorporating intellectual behavior in blockchain-based DTs, the functions of machines can be governed parallelly, and the major maintenance activities can be predicted effectively. This chapter explores the design of various frameworks using recent technologies. Some of them are DTs, blockchain, artificial intelligence, and the Internet of Things which drive resolution in the industry to minimize the governing cost of machines.

Blockchain-Based Digital Twins: Research Trends and Challenges.
A. Prasanth, Vandana Sharma, Rajesh Kumar Dhanaraj, Pethuru Raj (Eds.)

10.1 INTRODUCTION

Machine maintenance is vital in any industry or plant that utilizes mechanical parts. It helps industries meet production schedules and improve safety by preventing accidents and injuries due to faulty machinery. Regular maintenance is the key point for enhancing the performance level and longevity of machines. The major classification of maintenance includes proactive and reactive maintenance. In reactive maintenance, only broken machines will be brought into working condition once they are found to be repaired [1]. But we need to wait until something goes wrong with the machine.

Proactive maintenance is carried out in three forms: preventive, conditional-based, and predictive maintenance [2]. In preventive maintenance, the machinery is maintained at predetermined intervals without considering the current state of the machines, whether it is good or bad. Condition-based maintenance does not require preplanning, but the machines need to be monitored regularly. Once a change is sensed, the machines start to display problems that can be sensed by the sensor, and the maintenance process is carried out. The third kind of proactive maintenance, predictive maintenance, is considered to be more important than the other two. Predictive maintenance combines the strength of preventive and conditional-based maintenance.

The condition and performance of the machines are regularly monitored and tracked using the real-time data collected using the sensors. The need for maintenance is predicted by analyzing the machines' condition using this historical data. These predictions help the organization to prevent unexpected failure of the machines and hence avoid costly downtime. The Internet of Things (IoT) plays a major role in monitoring machines regularly [3]. With the advent of industrial automation and the technological boom, industries have shifted from condition-based maintenance to predictive maintenance.

The major benefits of predictive maintenance:

1. **Fewer Failures of Machinery:** Regular monitoring of the machine conditions and process systems helps to minimize failures by predicting the need for maintenance in advance.

2. **Increased Equipment Lifespan:** Prevention of failure reduces the severity of impairment and the dissemination of defects. It helps to improve the lifespan of machines.

3. **Effective Management/Decision Making:** The ability of predictive maintenance to use real-time data enables managers to make decisions on when to replace machinery.
4. **Increased Safety:** Early prediction of problems minimizes the possibility of fatal accidents and injuries.
5. **Cost Reduction:** As predictive maintenance predicts the health status of the machines and decides when the parts need to be replaced or repaired, the need for unnecessary inspection and maintenance can be avoided. Hence, the cost incurred due to this unnecessary maintenance can be eliminated.

The effectiveness of predictive maintenance relies on three major components which are depicted in Figure 10.1.

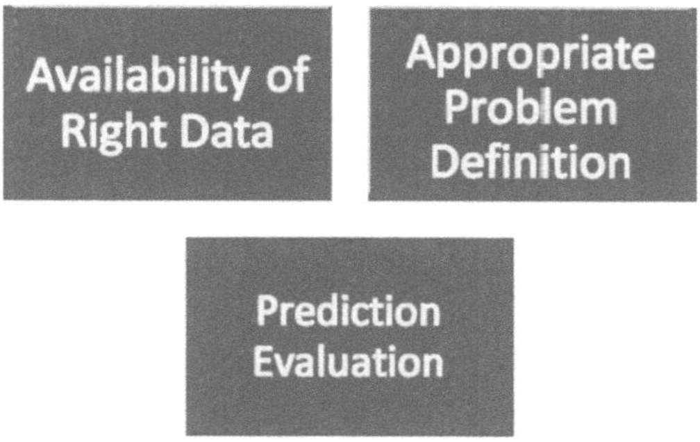

FIGURE 10.1 Components of predictive maintenance.

Predictive maintenance has taken one step further and become smart by leveraging the benefits of modern technologies like blockchain, edge computing, machine learning, and DT. The benefits and the role of these technologies are as in subsections.

10.1.1 BLOCKCHAIN FOR PREDICTIVE MAINTENANCE

Smart predictive maintenance relies on the real-time data collected from industrial machines to learn predictive models. The decentralized and immutable characteristics of blockchain technology draw the attention of researchers/industries and are considered to be an optimal solution for safeguarding the sensitive information collected from the industry in real-time for training the predictive maintenance model. Blockchain technology can minimize cybersecurity risks [4]. Most of the data related to the various kinds of plants, starting from power plants, manufacturing plants, refining plants, etc., is currently handled manually by the field workers. To optimize the functioning of these plants and to perform predictive maintenance, the data gathered from various IoT points needs to be digitized and shared securely in a decentralized manner.

For the predictive maintenance of physical equipment, the health status of individual machine subparts needs to be regularly monitored. Hence, the data of vital parameters which describe the current state of machine parts should be monitored. In the present time, most machines are constructed by assembling various parts manufactured by different industries. Hence, predictive maintenance seeks the involvement of different organizations, and it is required to ensure integrity and transparency in the parameter values to be shared among them.

In Blockchain, the blocks that carry the real data are stockpiled in a distributed manner [5]. Basically, the chain of blocks is warehoused in a distributed peer-to-peer network. Each block also contains the hash of its data value and the hash of the previous block. The cryptographic nature of this hash ensures the genuineness, transparency, and integrity of the data. The verification process adopted by the blockchain to permit the addition of new blocks in the chains creates trust among the organizers who share the data. Also, the consensus algorithm adopted in blockchain ensures that the data cannot be modified without the consensus of all the stakeholders or nodes in the network. This very basic characteristic of blockchain makes it suitable for DT-based predictive maintenance of the equipment that is assembled using parts manufactured by different vendors. The basic structure of the blockchain is given in Figure 10.2.

10.1.2 EDGE COMPUTING

Edge Computing in association with Blockchain has become a successful combination in providing a secure, distributed platform for IoT-based applications [6]. The data generated by the IoT-enabled machines is huge. Also, IoT-enabled machines do not have enough processing power and storage capability to handle this huge volume of data. Edge computing addresses this issue of IoT sensors by bringing computing and storage closer to the machines where the data is collected [7]. Also, the integration

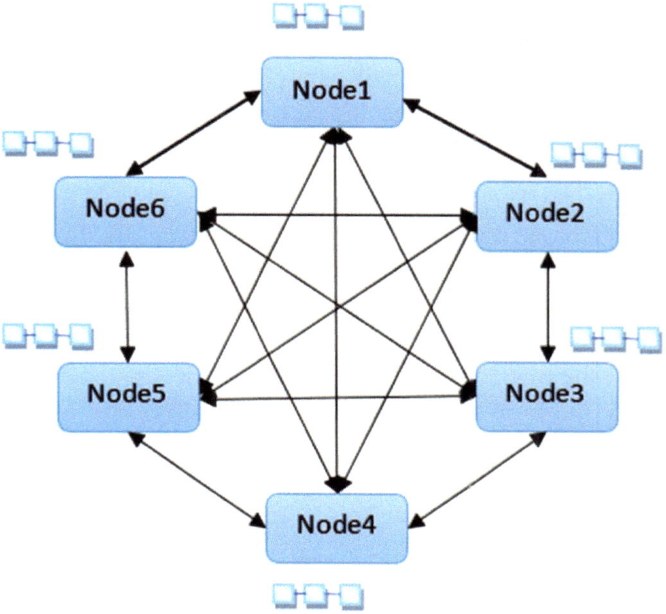

FIGURE 10.2 Structure of blockchain.

of IoT with edge computing provides a way for rapid data analysis. Enormous amounts of data from IoT and the need to transmit over the network result in bandwidth problems.

The utilization of bandwidth can be optimized with the support of edge computing by transmitting only the long-term storage to the processing center. The longer latency problem of traditional cloud architecture-based IoT applications is also addressed in edge computing by making the response time faster. In DT-based predictive maintenance, the data sensed from the machine end needs to be communicated to the cloud for the construction of blocks. Then the data from the block is fed into an analytical model to make decisions about maintenance. The communication of data from the origin end to cloud infrastructure faces two major challenges including bandwidth consumption and security problems.

1. **Bandwidth Consumption:** To make the DT imitate the functioning of the real-time machine and to do predictive maintenance analysis, the data sensed from the machines is communicated to the destination frequently. This frequent communication

consumes a large amount of communication bandwidth due to its huge size and frequency. This can be addressed by aggregating the continuous stream of data [8] and communicating only this aggregated data to the blockchain architecture over the cloud, which results in a notable reduction in the communication load.

2. **Security:** The vital reason for integrating blockchain technology in DT-based predictive maintenance is to ensure the security of the data common for multiple stakeholders who are the manufacturers of different parts of the machine. If the node from the cloud infrastructure is deputed with the responsibility of creating the block, then the raw data from the source end needs to be communicated over the network to the node. This makes the data exposed to security attacks. By bringing the block creation process close to the source end, the traveling of data in raw format over the network can be avoided, and hence security risk can be minimized.

The two challenges discussed above can be addressed using Aggregation and the Creation of blocks at the source end. But as mentioned already, IoT-enabled devices like sensor-fixed machines do not have enough computing power and storage. Hence, edge computing is considered a suitable solution for implementing these two processes. Figure 10.3 illustrates the architectural integration of IoT-enabled machines with Blockchain using edge computing.

10.1.3 MACHINE LEARNING MODEL FOR PREDICTIVE MAINTENANCE

Predictive maintenance indeed helps in determining the condition of industrial machines by analyzing the working status of individual parts and predicting when the machine parts require maintenance, and what type of maintenance to be given, like repairing or parts replacement. Implementation of a Machine Learning (ML)--based predictive model results in cost savings by avoiding unexpected downtime due to equipment failure. For predictive maintenance, data is collected over a period of time by continuously monitoring the state of machines in real-time. Besides, to facilitate the prediction of failure and maintenance, blockchain-based DTs of vital parts of the machines are constructed and data is extracted from these DTs.

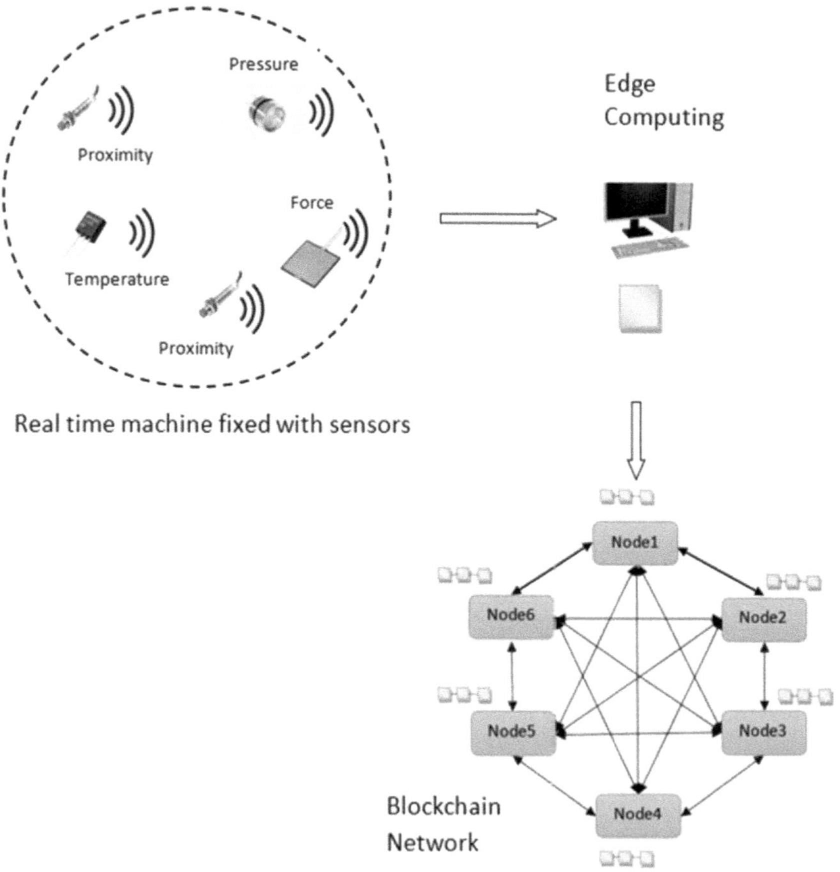

FIGURE 10.3 Integration of IoT-enabled machine with blockchain using edge computing.

This data is stored in a decentralized manner in blockchain in our proposed framework. The main goal is to find the patterns that assist in predicting and preventing machine failures.

10.1.3.1 ROLE OF ML

- Predictive models can be created to maximize the lifetime of equipment, functioning effectiveness, and uptime.
- Anomalies and failure patterns can be detected, and early warnings can be given.

There are multiple modeling techniques for predictive maintenance. Among them, the regression model can be used for predicting the remaining valuable lifetime. Failure can be predicted using classification models. Multiple types of failures can also be predicted by framing them as multi-class problems. In the classification model, prediction is done within a predefined time frame. In both these models, static and historical data are utilized for the prediction of failures and relationships among features. It is also used in modeling the degradation path [9]. The huge volume of data collected from real-time entities as well as from the DT helps the model to understand the relationship between characteristics of equipment during normal and failure times.

10.1.4 DEEP LEARNING TECHNIQUES FOR PREDICTIVE MAINTENANCE

Deep learning (DL) models perform better than traditional and statistical ML models in the area of predictive maintenance when adequate data is available. The term "DL" denotes neural networks, an artificial machine learning procedure inspired by the operation of the brain that goes beyond shallow 1 and shallow 2-hidden layer networks. Commonly used DL techniques for predictive maintenance are summarized below:

1. **Feed Forward Multilayer Perceptron (MLP):** A fully connected Artificial Neural Network (ANN) with three kinds of layers: input, hidden, and output. It can have more than one hidden layer. The model learns from the observations fed to the system by mapping the input features and target features [10].
2. **Convolutional Neural Network (CNN):** It is a class of neural networks that extracts features directly from data by applying convolutional filters and detects patterns in images. It is more effective in classifying sequential data like acoustic, time-series, and signal data. It is considered to be faster than the traditional learning model. But shows poor performance in modeling long-term dependencies [11].
3. **Recurrent Neural Network (RNN):** This backpropagation-based neural network can recognize patterns from sequential and time series data. This learning model helps in predicting the next possible scenario.

10.1.5 DIGITAL TWIN

The DT is considered to be a simulated depiction of real-time entities and has become an inevitable part of industries. DT replicates the operations of its physical counterparts. The DT can be of two kinds – static digital twin (SDT) and dynamic digital twin (DDT) [12]. SDT is a copy of the practical world entity and is not influenced by the data generated from its counterpart. But DDT can be paired with physical counterparts and remains updated using the data stream generated by sensory devices. DT can be used for building predictive maintenance models.

10.1.5.1 CHALLENGES IN THE CREATION OF DIGITAL TWIN

1. **Interoperability:** it is considered a primary issue. While establishing DDT, the information between a real-time entity and its virtual counterpart needs to be communicated in coordination without the support of the end user so that the DDT can replicate its counterpart. Predictive maintenance of multiple dependent machines requires the DTs to communicate and share information. Also, the DT which replicates the complex physical entity is made up of a hierarchy of smaller DTs. Communication among these building blocks of DT would be synchronized and coordinated. Interoperability plays a vital role in the parallel governance of the functioning of these DT building blocks.
2. **Storage:** The stream of data communicated between DT and the physical entity, between DTs has to be stored in one place. It should be restructured to reflect the present standing of the corporal entity. Both archival and conceptual information in the storage should be available to the learning system for analysis in time. The more data, the better the performance of DT. Hence, the storage should be able to handle huge volumes of data. Also, it is essential to remove redundancy in the data stored in DT and avoid bottlenecks.
3. **Transparent Communication:** The DT constructed from multiple contributors replicates the data from multiple physical entities. Hence, ensuring transparency in the data being transmitted among

the contributors is another challenging issue to be considered while designing the DT.

4. **Information Technology (IT) and Network Infrastructure:** For successful function, DT requires a combination of networks that can handle the transportation of huge volumes of data with high speed and support of IoT devices. Also, it is necessary to preprocess the data and do precomputing before sending the data to the digital counterpart. This can be facilitated by bringing computation and storage facilities close to the source environment using Edge computing, a distributed IT infrastructure.
5. **Security:** The creation and functionality of a DT depend on the huge volume of sensitive data collected from the physical counterpart. The integrity, confidentiality, and transparency of this sensitive data need to be ensured.

By considering the importance of predictive maintenance machines in the industry and the reimbursements of various technologies, we propose to develop a framework for a cost-efficient blockchain-based DT for the predictive maintenance of machines using machine learning.

The chapter is structured as follows: Section 10.2 addresses the various real-time issues related to DTs, blockchain, and ML. Section 10.3 illustrates the various stages in the construction of the proposed framework and its architecture. Concluding notes are specified in Section 10.4.

10.2 EXISTING SYSTEMS

In this section, the various frameworks proposed using the DT and blockchain technology to address the problems in various real-time applications like Vehicular Ad hoc NETwork (VANET), Industry 4.0, etc., are discussed. In Ref. [13], a DT model has been proposed for VANET with improved secure information transformation using Blockchain and low average delay time. It used digital signatures to verify the correctness of the traffic information passed to the Roadside unit. The signature is generated using SHA256 and security is strengthened by adopting blockchain technology.

Shamir's Secret Sharing (SSS) algorithm [14] is adopted by this scheme for distributing the message among unknown opponents. The message is sliced into an 's' number of fragments. The protecting strength is based on the two-threshold values t,s, where 't' refers to the minimum count of vehicles to participate in the cooperation process for the recovery of messages from the fragments, and 's' refers to the total count of slices generated by the SSS. It is further strengthened by reducing the risk of communicating with harmful nodes. Harmful nodes are identified by calculating the direct trust using Bayesian estimation and indirect trust. Hence, trust and fragmentation-based secure data transmission using Blockchain technology minimizes the effect of change in vehicle speed on the performance of the algorithm.

Kumar et al. [15] projected research work to facilitate the data analysis of the critical processes in the Industrial Internet of Things. The two major security elements – integrity and authenticity, are assured by adopting a blockchain-based data communication scheme. Also, the authors projected a Deep Learning-based algorithm for detecting intrusion in blockchain by learning the long-distance feature.

In Ref. [16], the authors proposed a Machine-as-a-Service model, which facilitates the industries to rent their equipment with a variety of sensors attached to the customers, monitor those rented machines continuously, and implement maintenance work to avoid unexpected downtime. The proposed model leverages the benefits of the IoT and blockchain. The IoT-generated data is securely monitored using a decentralized ledger. In Ref. [17], the authors proposed a use case for making the machine smart to address the problems of Industry 4.0. The predictive maintenance model is developed using machine learning which facilitates the machine to predict its remaining lifetime. The authors presented a DT for monitoring the smart machine.

In Ref. [18], the authors proposed a framework for monitoring the operational behavior and health of semiconductor lasers used in optical communication. A Machine Learning-based model is developed for predictive maintenance. To predict possible degradation in performance, a recurrent neural network-based learning model called gated recurrent unit (GRU) is adopted. GRU can retain information for a certain duration using update gates and reset gates, allowing the model to refine the output by adjusting the flow of information. Subsequently, an unsupervised reduction-based model that learns by clustering data is used to detect performance degradation in the laser. Another learning model utilizing an attention-based neural network is employed to identify the remaining healthy lifetime.

In Ref. [19], the authors discussed how to increase the sustainability of IoT devices. Lack of resources and ineffective management have seriously hampered the growth of sustainability. The DT technology makes it easier for the IoT and its digital counterparts to interact. High-fidelity DT replicas, however, increase the need for effective data flows, which are constrained by practical factors like data gathering methods and energy utilization. The authors proposed a workable method for gathering and managing data to create a DT. In this method, lots of IoT devices upload data packets to the data brokers, which are called agents. The test lies yet to be determined between getting through information assortment and the data misfortune related to the lifeless information. The objective of this chapter is to guarantee both sustainable information and energy while simultaneously optimizing the data integrity and disclosing latency metrics. Moreover, a shareable and economical blockchain-based computerized twin administration design is put forward, which doesn't depend on information exchanges with a single unified server.

In Ref. [20], the authors proposed a framework for data management and security. Sensory data is essential for crucial decision-making processes in industrial processes. An infrastructure that can guarantee the data's trustworthiness is necessary for obtaining useful insights from the gathered data. A framework for data management is developed for the Industrial Internet of Things (IIoT) based on blockchain technology. After the data gathered from reliable origins has been logged in the blockchain, machine processing cycle events can be given into data-driven systems for process monitoring, diagnostics, and optimized control. DTs are used along with data analysis mechanisms to draw intelligent interpretations from the data by locating flaws and recommending preventative measures before crucial events.

Applications that use sixth-generation (6G) wireless networks for the Internet of Everything (IoE) have a variety of requirements for latency, dependability, data rate, and user-specific performance measures. As a result, to run and enhance the 6G wireless network and other related IoE services, it is necessary to create a new framework. On DTs, a new architecture for 6G could be established with the help of related algorithms, communication technologies, computing platforms, privacy mechanisms, and the virtual depiction of the physical 6G system (e.g., blockchain). In Ref. [21], the author proposed the essential design specifications for utilizing a DT to enable 6G. Architectural trends and elements including twins based on edge-cloud, cloud, and edge are discussed. Also, a comparative description of different twins was done.

The potential for DT technology to be used in business and industry has recently brought it to the forefront. In DT atmospheres, information produced by physical assets is sent to a distant server, which runs imitations using DTs in a virtual environment. The owner of the data or other users can then receive the simulation's findings. The use of DT technologies in the real world is not without difficulties. Identifying secure sharing methods for simulation data and real-time data are two of the major issues. Sensitive information from data owners may be included in the data created by physical assets, and privacy issues could arise if data were to be disclosed to an enemy. To optimize the availability of DT data, it should also be thought about sharing data with other data consumers. The authors in Ref. [22] suggest a system paradigm for the safe exchange of DT data to address these problems. The suggested system concept makes use of blockchain technology for data verifiability and cloud computing for effective data sharing. Communication plans for the suggested paradigm to ensure data security and privacy protection through wireless channels are provided.

In Ref. [23], the authors imagine how blockchain may change and alter DTs to enable safe production that ensures transparency, adherence, legitimacy, integrity, and security. Based on important criteria, the DT literature (for example: DT tiers, design stages, use cases in the industry, important goals, supporting technology, and fundamental applications) is categorized. Blockchain technology can improve the idea of DTs by ensuring openness, decentralized storage space, data integrity, and peer-to-peer interaction in industrial sectors. The DTs play a critical role in preserving and managing Industry 4.0 since they assist in visualizing designs in three dimensions, conducting testing and simulations virtually before the construction of any physical component. Because DTs can be utilized for configuring, observing, diagnosing, and prognosticating, it is anticipated that they will become more common in the near future.

A Systematic Literature Review on predictive maintenance, which is considered as a method for developing a more sustainable, secure, and successful industry, is provided in Ref. [24]. Lack of failure data is one of the biggest obstacles to the development of predictive maintenance systems since machines are routinely repaired before they fail. The predictive maintenance algorithm can employ data from DT, which offers a depiction of the physical world equipment in real-time and creates information about asset depreciation. Since 2018, there has been an upsurge in the scholarly works on the use of DTs for condition-based preservation, signifying the need for an extensive investigation. To open the door for more research, this study gathers and synthesizes works that concentrate on predictive maintenance employing DTs.

In a manufacturing organization, predictive maintenance is strategically important to preserve high quality in production and prevent unplanned stoppages. To plan preservation cycles and maximize productivity in this situation, the future health status of the machines must be predicted. The suggested method is based on the application of Electronic Design Automation methods that have been translated from the electronic field to the field of manufacturing lines. In Ref. [25], the authors developed a broad framework based on the Electronic Design Automation methodology that enables the establishment of a maintenance strategy through the analysis of sensor data. Predictive maintenance and online monitoring are just a couple of the execution modes it possesses. The methodology's usefulness has been demonstrated through application to a mechanical transmission system.

The aforementioned survey reveals that the DT plays a vital role in learning and collecting more vital information about the real-time entity which can be used effectively by the analysis model. The dataset supplied into the machine learning model is the only factor that influences how well it performs. Finding a suitable dataset that provides maximum knowledge of the learning model is a challenging task. DT addresses this issue by generating a stream of data that describes the state of the real-time entity being imitated. Blockchain technology is incorporated into the existing system to address the security issues related to the sensitive information gathered from the DT. The survey also discloses the importance of predictive maintenance in the industry. Hence, this chapter proposes a framework for predictive maintenance by leveraging the benefits of blockchain, DT, and DL.

10.3 FRAMEWORK OF BLOCKCHAIN-BASED DIGITAL TWIN FOR PREDICTIVE MAINTENANCE

A framework of Blockchain-based DTs for the predictive maintenance of industrial machines has been proposed. The various stages in the development of the framework are elaborated.

10.3.1 PRE-DESIGN PHASE

In this phase, information required for constructing the DT so that the DT will imitate the functionalities of real-time physical machines is identified. The required information is classified into functional information and operational influencing information. The functional information provides insight into the sequence of operations to be replicated by the DT for predictive maintenance like parts of the equipment, the functionality of each part, physical characteristics of each part, operational sequence, logical and physical connection among the parts, etc. The operational influencing information refers to the factors that affect the performance of the machines and external factors that influence the function of machines like temperature, oil, number of hours being operated, humidity, etc.

1. **Virtua Entity (VE):** Based on the information identified in the pre-design phase, virtual entities corresponding to their physical counterparts and additional virtual entities required so that the functionality of physical parts can be replicated in the DT need to be identified.

2. **Hierarchy Relation of VE:** The collection of virtual entities and the hierarchical relationship among them creates the virtual representation of the physical entity to be imitated by the DT. Hence, based on the operational sequence of the physical entity to be replicated, the hierarchy of the VEs has to be created by establishing their functional relationships. The hierarchy of VE will form the DT of the sub-physical part of the complex machine to be replicated.

   The general architecture of the sample physical equipment to be replicated is illustrated in Figure 10.4. Part 1, Part 2, ..., and Part 7 refer to the sub-parts of the equipment. The construction of a DT which imitates the function of Part 1 with n number of virtual entities is given in Figure 10.5. For example, if the washing machine is the equipment to be imitated by the DT, then the various subparts are the controlling panel, water level controller, drain valve controller, inlet valve controller, motor control, etc.

3. **Virtual Connection of DT:** To imitate the whole function of complex real-time entities or machines, it is required to construct

multiple DTs. Each DT constructed using the hierarchy of VEs will form the building blocks of the complex digital twin. It can be constructed by identifying the suitable hierarchical relationship and by establishing a virtual connection between building block DTs. Figure 10.6 shows the virtual connection among the sample DT building blocks.

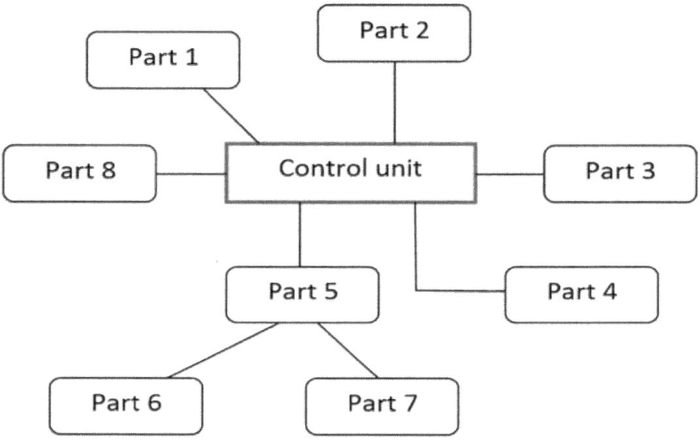

FIGURE 10.4 Architecture of sample physical equipment.

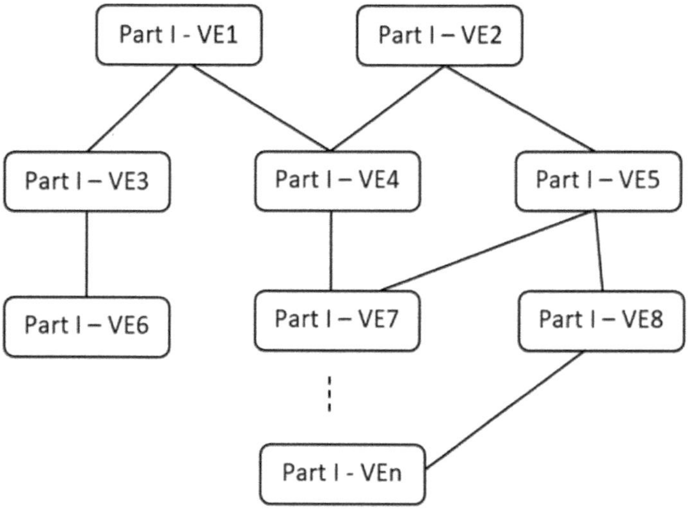

FIGURE 10.5 Digital twin of the physical part I.

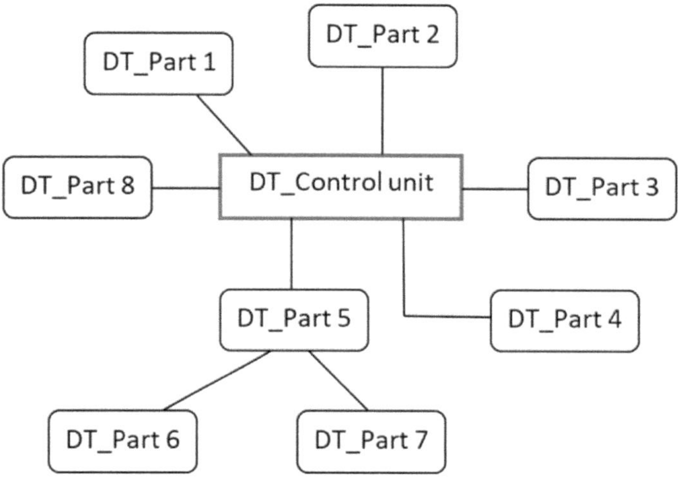

FIGURE 10.6 Digital twin of physical equipment.

10.3.2 ENCODING PHASE

Sensors are fixed to collect the information that replicates the current state of the machines. The collected data are encoded into a digital form that is to be communicated to the DT.

10.3.3 BLOCKCHAIN CONSTRUCTION PHASE

Integration of DTs with blockchain guarantees vital features like data integrity, tracking, immutability, and security. All the data related to the construction, policies, rules, and operation of DT can be maintained securely in Blockchain. The architecture of blockchain is classified into three categories private, public, and consortium blockchain. Access permission, consensus determination, and level of immutability are determined based on the kind of architecture adopted. In our proposed framework for DT-based predictive maintenance, the status information of different machine parts needs to be tracked by different organizations. Hence, consortium-based architecture is considered to be more suitable.

Consortium blockchain is a hybrid form of private and public blockchain. The sensor data which exhibits the current status of the real-time machine parts and DT need to be maintained in the blockchain for predictive maintenance. It can ensure data accountability, security, and transparency among the members of different organizations who are involved in the predictive maintenance process. In the proposed framework, two different blockchains are constructed in order to improve the efficiency of the learning model: Digital Twin Blockchain (DTBC) and Real-time Machine Blockchain (RMBC). RMBC carries a chain of blocks with data obtained from the real-time physical entity to be maintained whereas DTBC includes blocks that carry information obtained from the DT.

10.3.3.1 CONSTRUCTION OF RMBC

In this section, the steps for creating a blockchain are discussed.

- **Step 1:** Creation of Aggregate Data Structure: Each block in the blockchain carries the data used for tracking the state of individual parts of the machine to be maintained. The data stream generated by the sensing devices that are fixed in the machine equipment is huge. To minimize the cost of storage and communication, the number of blocks in the blockchain is to be reduced. This can be achieved by aggregating the stream of data which is generated by the sensing devices. For each major part, the virtual entities used to describe the state of the part are collected by edge computing devices that act as aggregators. The aggregator computes the aggregate representation of all the virtual entities. These aggregates are grouped based on the parts to which they belong.

  Depending on the block size, either a separate block for each part can be created, or aggregates of multiple parts can be stored in a single block along with part identification and time. This information helps to track the status of machine parts at different times. The possible aggregate functions are SUMMATION, AVERAGE, COUNT, MAXIMUM, or MINIMUM. The description of the aggregation function is provided in Table 10.1. Depending on the kind of virtual entity and the parts to which it belongs, different aggregation functions can be applied. Hence, information regarding the kind of aggregate function has to be stored along with the aggregate. Figure 10.7 exhibits the structure of data which describes the state of a machine part to be stored in the chain block. Let the values that occurred during the particular time intervals be $v_1, v_2, \ldots, v_j$.

TABLE 10.1 Aggregation Function

Aggregate Function	Description
SUMMATION	$\sum_{i=1}^{n} v_i$
AVERAGE	$\frac{\sum_{i=1}^{j} v_i}{j}$
MAXIMUM	$maximum\ (v_1,v_2,\ldots,v_{j'})$
MINIMUM	$mininmum\ (v_1,v_2,\ldots,v_{i'})$
COUNT	'j'

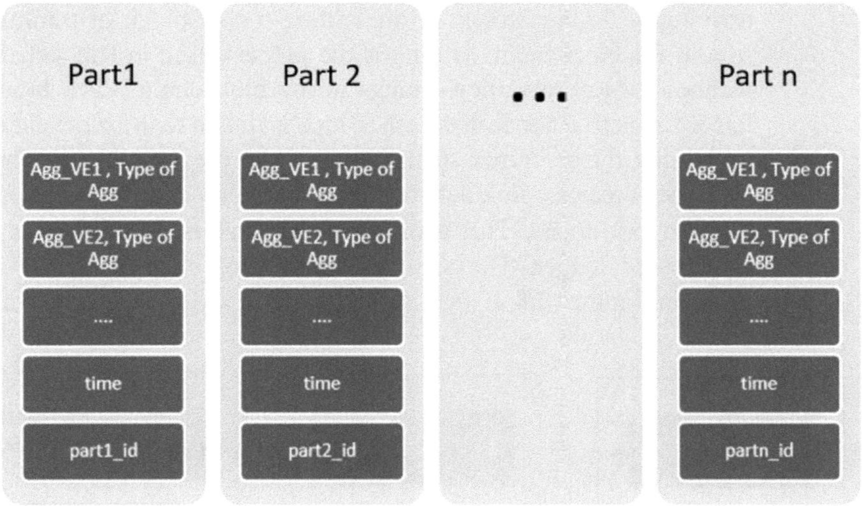

FIGURE 10.7 Structure of data about each part.

By considering the volume of data generated from multiple sensing devices and resource constraint problems associated with the sensing device, edge computing is adopted for storing the enormous data stream belonging to a particular time duration and computing the aggregate value.

- **Step 2:** Encryption of Aggregate Data Structure: After the creation of an aggregate data structure, it should be added to the blockchain. From the edge computing device, the aggregate has to be broadcast to all the nodes in the blockchain for the mining process. Hence, it is required to ensure the security of the data being broadcast. This can be achieved by applying cryptography and converting the data aggregate into a non-understandable format.

- **Step 3:** Block Creation and Addition: To ensure reliability and confidence among the stakeholders about the data maintained in the blockchain, consensus algorithms are mainly designed. It enables the computer nodes to agree on the state of the distributed ledger. The consensus mechanism is the way to reach an agreement between validators on who has the right to add a new block and what transaction. Proof of work consensus algorithm over the Ethereum network is adopted in the proposed framework for arriving at the agreement while adding a new block of transaction in the blockchain. Mining is the process used in Blockchain technology to create new blocks in the blockchain. Each block has a distinctive nonce and hash. Block is linked to the blockchain by adding the reference of the hash of the previous block. This reference creates a link between blocks that are distributed across the network nodes. This distributed database is referred to as a Distributed ledger. The sample structure of chained blocks is given in Figure 10.8.

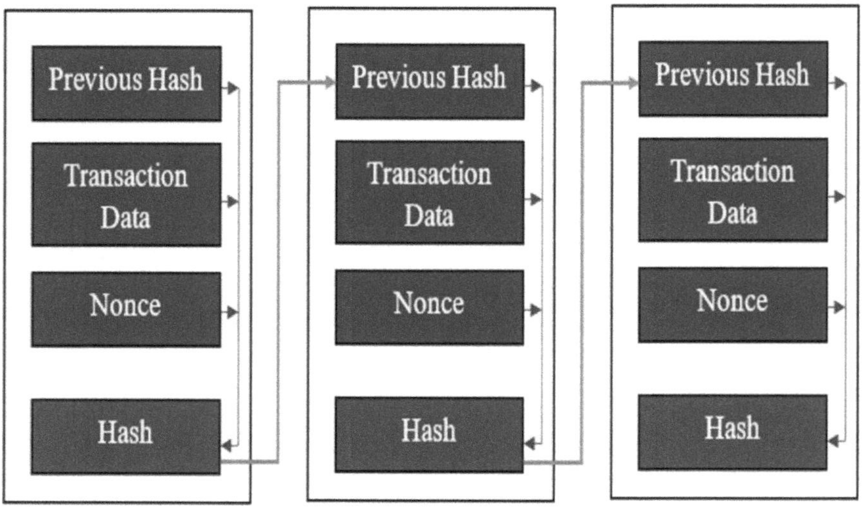

FIGURE 10.8 Blockchain structure.

Finding the suitable nonce that generates the acceptable hash is a complex math problem. To compute this complicated mathematical puzzle, the miner requires large computation power. Also, to make the hash acceptable and hence the block valid, the hash has to meet certain criteria like some number of leading zeros. For this purpose, the field nonce (a number used once) is utilized. Finding the best nonce is the miner's complicated mining task. Independent of the number of times the hash function is applied, it should generate the same hash for the same input. The length of the hash is fixed. The process of adding a new block is illustrated in Figure 10.9.

The major applications of the hash function in Blockchain are listed as follows:

1. **Integrity:** By keeping the root hash inside the block header, the integrity of the block header as well as the transactions can be protected. Because no two Merkle trees with the same root hash are impracticable.
2. **Authentication:** By generating the hash using the node-specific secret key, it can act as a digital signature which ensures the authentication of the blocks.
3. **Confidentiality:** The transactional data communicated to the blockchain network from the edge computing device is encrypted. Adopting asymmetric privacy homomorphism which allows the computation over the cipher data ensures end-to-end confidentiality.
4. **Chain of Blocks:** The hash is generated by applying the hash algorithm over the hash of the previous block, transaction data, and nonce. The inclusion of the previous block hash creates a virtual chain link between blocks.

   The most commonly used hash functions in blockchain technology are:

   i. **SHA-256 Algorithm:** This algorithm deterministically generates the hash output for the given amount of data. The output hash is 64 characters long.

   ii. **ETHASH Algorithm:** It is an algorithm responsible for controlling the mining process of Ethereum Proof of Work.

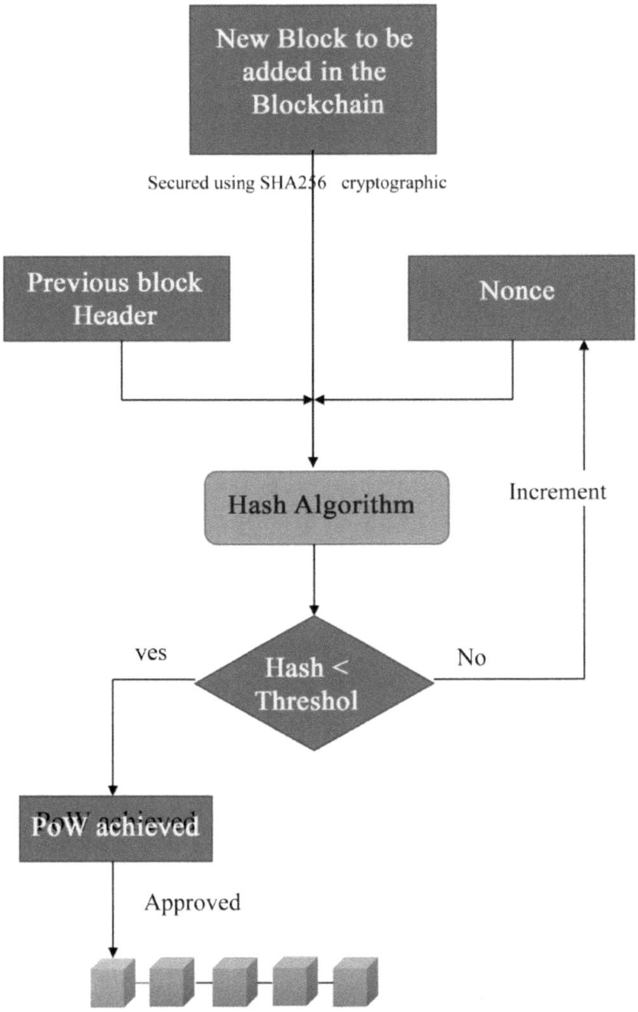

FIGURE 10.9 Addition of a new block to the blockchain.

The utilization of DAG for mining makes it a memory-hard function. The memory cost is so high that mining is possible with a GPU (Graphical Processing Unit).

iii. **Scrypt Algorithm:** It is considered to be an optimized hashing algorithm as it can support huge volumes of data, guarantee integrity based on cryptography, and support the recovery of data.

10.3.3.2 DIGITAL TWIN BLOCKCHAIN

The DTs on each physical part of real-time machines based on Blockchain are made up of a blockchain network of interconnected DTs. It is a network of multiple DTs connected through the blockchain. The DTBC ledger stores information about the status of all virtual entities of DT, the owner of DT, the recording time at which status is being recorded, the synchronous time at which the DT is being synchronized with its physical counterpart, and data being exchanged between the DTs. Figure 10.10 illustrates the data structure of the DTBC ledger.

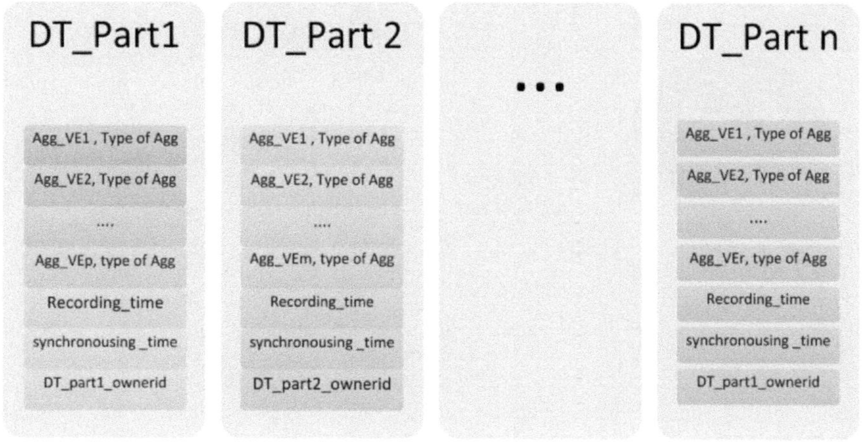

FIGURE 10.10 The data structure of DT_Parts.

The blockchain network of the DT is given in Figure 10.11. Blocks carrying the transaction data of each DT_Parts are created and added to the cloud-based blockchain network proof of work Ethereum-based consensus algorithm using the steps described for the construction of RMBC.

10.3.4 CONSTRUCTION OF DEEP LEARNING MODEL

To model the temporal relationship of sequential data collected from real-time IoT-enabled machines and its digital counterpart that is the collections of DTs, the RNN-based learning model is considered for the proposed framework. To predict the failures and recognize the patterns that require timely maintenance, it is required to model the longer-time dependencies from the data that describe the status of machine parts collected at different periods. To predict the abnormality in the operational behavior of machines, it is needed to identify the difference between expected target characteristics and the current behaviors of the machine.

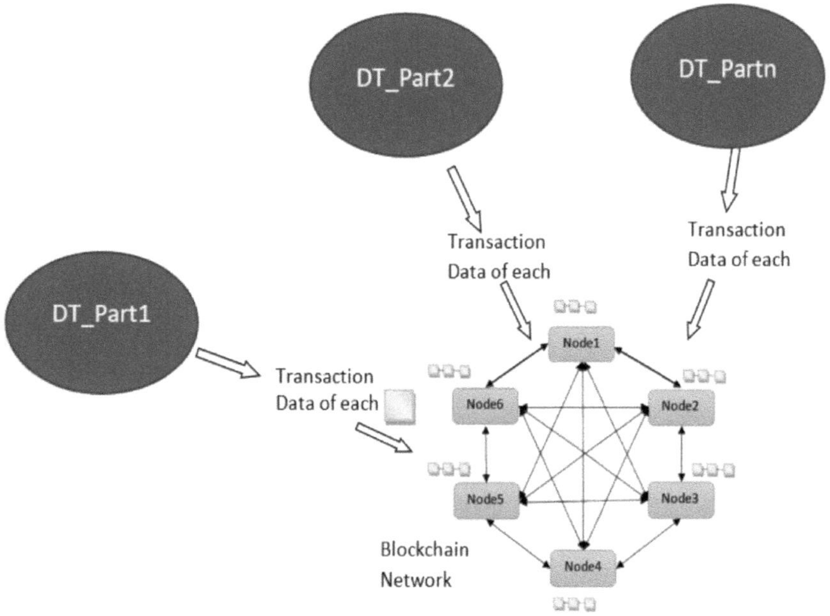

FIGURE 10.11 Blockchain network of digital twin.

This is supported in the proposed framework through the two kinds of blockchain RMBC and DTBC. The DTBC provides the expected behavior of the machine. Another requirement for predictive maintenance is finding the relationship between the sequence of state information. Let s_t be the state of the machineat time '*t*' and s_{t-1} be the state of the machineat time '*t*-1.' While analyzing the s_t for predicting the maintenance requirement or probability of failure, the knowledge of the machine state at a time '*t*-1 ,' i.e., s_{t-1} is also to be considered. The Preciseness of the prediction task can be improved by also considering previous 'n' state information like s_{t-1},s_{t-2}, s_{t-3}...s_{t-n}. Hence, the Long Short-Term Memory (LSTM) neural network is considered to be a suitable learning model for the proposed predictive maintenance.

The LSTM can do analysis and classification of current input using the knowledge acquired from the past data. The LSTM can decide which information is worth remembering or not. The architecture of LSTM has four gates to support learning based on long-term dependencies. The gates decide what to learn, forget, remember, and use. The Learning gate takes the current state C_t and previous short-term memory SM_{t-1} as input. It decides which information is relevant and keeps only those as learning for prediction. The operation of the learning gate (lg) using the weight matrix (M_n) is given below:

$$Learning\ Matrix\ (L_t) = tanh\,tanh(M_n[SM_{t-1,}S_t]) + bias_n \quad (1)$$

$$Ignoring\ factor\,(ig_t) = (M_n[SM_{t-1,}S_t]) + bias_n \quad (2)$$

$$lg_{output} = L_t.ig_t \quad (3)$$

The Forgetting gate takes the long-term memory (SM_t) as input and decides what to be forgotten and what to be remembered as follows:

$$Ignoring\ factor\,(ig_t) = (M_n[SM_{t-1,}S_t]) + bias_n \quad (4)$$

$$LM_t = LM_{t-1}.((M_n[SM_{t-1,}S_t]) + bias_n \quad (5)$$

The output of the remembering gate is produced by adding the output of the learning gate and the forgetting gate. The use gate combines vital information from the previous long-term memory and short-term memory to produce the short-term memory for the next cell. Once the data from the RMBC is fed into the LSTM-based learning model, the type of failure is predicted. Prediction accuracy is improved with the help of DTBC. The architecture for the proposed framework twin is given in Figure 10.12.

10.3.4.1 CHALLENGES OF BLOCKCHAIN-BASED DIGITAL TWIN PREDICTIVE MAINTENANCE

- Reusability of PM model: The major challenge of predictive maintenance is the variability in the data and behavior of industrial machines. This is applicable for machines working under the same environmental features, making the reusability of models difficult for different machines.
- Though the LSTM adopted in the proposed framework can address the problem of vanishing gradient, which is underfitting, by replacing the sigmoid function in the hidden layer with ReLU, it converges quickly and suffers from extensive overfitting.

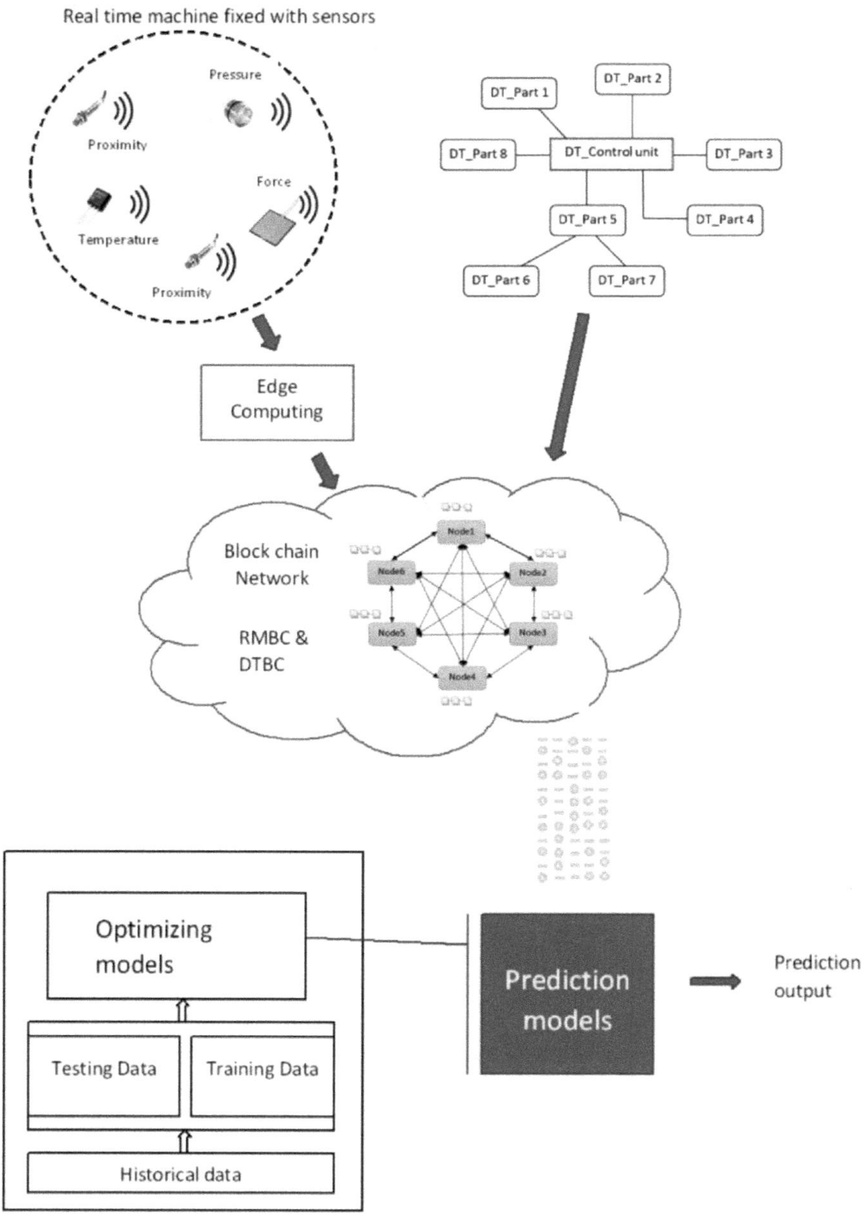

FIGURE 10.12 The architecture of blockchain-based digital twin for predictive maintenance.

10.4 CONCLUSION

In this chapter, a novel framework for predictive maintenance of the IoT-enabled industry was proposed by leveraging the benefits of edge computing, DT, and blockchain technologies. It utilized the LSTM for feature extraction and predicted the major maintenance activities like possible time and type of failure of different machine equipment and which parts of the machine require replacement or regular preventive maintenance activities at what time. The effectiveness of prediction functions was improved by managing the DTs of multiple parts of the machine. Parallel governance of multiple DTs provides a comprehensive insight into the functionalities of machines. The data collected from the DT enables the framework to identify the difference between expected target characteristics and the current behaviors of the machine. The incorporation of blockchain technology ensures the security of data of different machine parts and allows the smooth sharing of it among diverse stakeholders.

KEYWORDS

- **blockchain**
- **deep learning**
- **digital twin**
- **edge computing**
- **industrial machines**
- **long-term memory**
- **machine learning**
- **novel framework**
- **predictive maintenance**
- **stakeholders**

REFERENCES

1. Zwolinska, B., & Wiercioch, J. (2022). Selection of maintenance strategies for machines in a series-parallel system. *Sustainability, 14*(19), 11953.
2. Sharma, J., Mittal, M. L., & Soni, G. (2022). Condition-based maintenance using machine learning and role of interpretability: A review. *International Journal of System Assurance Engineering and Management, 15*(2).

3. Lavanya, S., Prasanth, A., & Jayachitra, S. (2021). A Tuned classification approach for efficient heterogeneous fault diagnosis in IoT-enabled WSN applications. *Measurement, 183*, 1–23.
4. Guo, H., & Yu, X. (2022). A survey on Blockchain technology and its security. *Blockchain: Research and Applications, 3*(2).
5. Thang, P., Trang, T., & Tai, B. (2022). Multimedia privacy, security, and protection within the blockchain: A review. *Proceedings of 5th IEEE International Conference on Contemporary Computing and Informatics*, 1171–1176.
6. Chandana, K. P., Samaniego, S. L., & Chaudhary, S. S. C. (2022). Intelligent mobile edge computing integrated with blockchain security analysis for millimeter-wave communication. *International Journal of Communication Networks and Information Security (IJCNIS), 14*(3), 110–122.
7. Hassan, N., Gilani, S., & Ahmed, E. (2018). The role of edge computing in the internet of things. *IEEE Communications Magazine, 56*(11), 110–115.
8. Vinodha, D., & MaryAnita, E. A. (2018). Secure data aggregation techniques for wireless sensor networks: A review. *Archives of Computational Methods in Engineering, 26*(8), 1134–3060.
9. Florian, E., Sgarbossa, F., & Zennaro, I. (2021). Machine learning-based predictive maintenance: A cost-oriented model for implementation. *International Journal of Production Economics, 236*, 108114.
10. Werbos, P. J. (2015). Applications of advances in nonlinear sensitivity analysis. *System Modelling and Optimization, 38*, 762–770.
11. Serradilla, O., Zugasti, E., & Rodriguez, J. (2022). Deep learning models for predictive maintenance: A survey, comparison, challenges, and prospect. *Applied Intelligence, 52*, 10934–10964.
12. Hasan, H. R., Salah, K., & Jayaraman, R. (2020). A Blockchain-Based Approach for the Creation of Digital Twins. *IEEE Access, 8*, 34113–34126.
13. Feng, H., Chen, D., & Lv, Z. (2022). Blockchain in digital twins-based vehicle management in VANETs. *IEEE Transactions on Intelligent Transportation Systems, 23*(20), 19613–19623.
14. Adi, S. (1979). How to share a secret. *Communications of the ACM, 22*(11), 612–613.
15. Kumar, P., Kumar, R., & Kumar, A. (2022). Blockchain and deep learning for secure communication in digital twin-empowered industrial IoT network. *IEEE Transactions on Network Science and Engineering*, 1–13.
16. Tran, V. H., Lenssens, B., & Kassab, A. (2022). Machine-as-a-Service: Blockchain-based management and maintenance of industrial appliances. *Engineering Reports*, 1–22.
17. Hichri, B., Driate, A., & Borghesi, A. (2022). Predictive maintenance based on machine learning model. *Artificial Intelligence Applications and Innovations, 647.*
18. Abdelli, K., Grieber, H., & Pachnicke, S. (2022). A machine learning-based framework for predictive maintenance of semiconductor lasers for optical communication. *Journal of Lightwave Technology, 40*(14), 4698–4708.
19. Wang, C., Cai, Z., & Li, Y. (2022). Sustainable Blockchain-based Digital Twin Management Architecture for IoT Devices. *IEEE Internet of Things Journal, 10*(8), 6535–6548.

20. Suhail, S., Hussain, R., & Jurdak, R. (2022). Trustworthy Digital Twins in the Industrial Internet of Things with Blockchain. *IEEE Internet Computing, 26*(3), 58–67.
21. Khan, L. U., Saad, W., & Niyato, D. (2022). Digital-Twin-Enabled 6G: Vision, Architectural Trends, and Future Directions. *IEEE Communications Magazine, 60*(1), 74–80.
22. Son, S., Kwon, D., & Lee, J. (2022). On the design of a privacy-preserving communication scheme for cloud-based digital twin environments using Blockchain. *IEEE Access, 10*, 75365–75375.
23. Yaqoob, I., Salah, K., & Uddin, M. (2020). Blockchain for Digital Twins: Recent advances and future research challenges. *IEEE Network, 34*(5), 290–298.
24. Dinter, R. V., Tekinerdogan, B., & Catal, C. (2022) Predictive maintenance using digital twins: A systematic literature review. *Information and Software Technology, 151*.
25. Centomo, S., Dall'Ora, N., & Fummi, F. (2020). The design of a digital twin for predictive maintenance. *25th IEEE International Conference on Emerging Technologies and Factory Automation (ETFA)*, Vienna, Austria, 1781–1788.

CHAPTER 11

Blockchain-Based Digital Twin for Agricultural Field Management

S. MAHESWARAN,[1] T. ARUNKUMAR,[2] R. D. GOMATHI,[3] P. RITHIKA,[1] I. MOHAMMED SHAFIQ,[1] and S. NANDITA[1]

[1]*Department of Electronics and Communication Engineering, Kongu Engineering College, Perundurai, Tamil Nadu, India*

[2]*Department of Computer Science Engineering, Kongu Engineering College, Perundurai, Tamil Nadu, India*

[3]*Department of English, Kongu Engineering College, Perundurai, Tamil Nadu, India*

ABSTRACT

Agriculture is the major source of the world's food and fabrics. Increasing population and globalization affect the demand for food. So, there is a need to develop cost-effective technologies to improve agriculture. One such technology is the digital twin. This twin technology is the computer-generated imaginary illustration of the factual article in the computer-generated imaginary world. A digital twin is a virtual replica designed to monitor, detect, prevent, predict, and customize using real-time data. In agriculture, the digital twin acquires data using sensors, satellites, drones, historical soil information, and soil probes. From observations, first, it creates a database using API data, live sensor data, and manual records. Then it processes the data and provides historical data as output. It comments on feedback, and errors, predicts future conditions, and suggests further improvement. It provides photos, digitized paper maps, farm data, moisture content,

Blockchain-Based Digital Twins: Research Trends and Challenges.
A. Prasanth, Vandana Sharma, Rajesh Kumar Dhanaraj, Pethuru Raj (Eds.)

weather, and crop properties. Furthermore, the specified data are stored using blockchain technology where the blockchain is a chain of blocks that contain information. It is used for computing and to store transactions and records. This chapter explores the usage of Blockchain and digital twins in the agriculture sector. It also deals with the sensing, computing, and communicating models for various agriculture applications. Finally, it enumerates the virtual model for healthy crops and non-healthy crops in the agriculture field.

11.1 INTRODUCTION

Most of the countries are agricultural-based nations. Agriculture is very important because it is the source of our life. In fact, in India, 68% of the people are based on income from agriculture and allied sectors. It is the main source of human food. Even small-scale productions and cottage businesses depend on farming production centers. Not only for food but also in our daily lives, all the necessary products are produced by agriculture (meaning all products' raw materials are from agricultural plants and trees). Therefore, in the current situation, the population is increasing drastically. Similarly, with the increase in population, pollution is also increasing. The nutrient level in the soil has decreased significantly, and the soil is heavily affected by the chemicals entering the rivers and lakes [1].

Due to the decreased nutrient level, crops are being affected at an early stage. After the nutrients, crops are further affected by pests and fungi at a higher rate. Even though farmers use pesticides and fungicides to control them, the crops still get affected. Moreover, farmers cannot identify all these issues at an early stage. Therefore, there is a need for automation to identify and address these issues at the initial stages. Looking at automation in agriculture, there are many innovations in management [2]. However, all these advancements face a common challenge, which is the collection, storage, and processing of data for management.

As agricultural fields contain millions of crops, the data collected for further processing is extensive. There is no backend option to retrieve this data for future use. Therefore, we decided to create a blockchain integrated with big data automation to monitor the crops, along with Artificial Intelligence (AI) to predict diseases at an early stage [3]. In the blockchain component, the data collected from the automation process is stored in the cloud for easy access at any time and for seamless transmission to the AI for evaluation metrics. Big data is utilized to store the large data extracted from the blockchain and real-time data for analysis. By using this data, the AI predicts diseases and provides information to the digital twin to visualize the real-time condition of the specific crop with the location indicated, as shown in Figure 11.1.

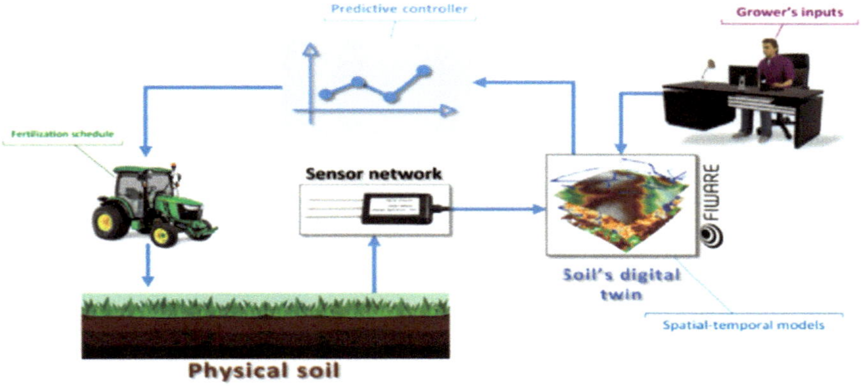

FIGURE 11.1 Outline of the entire process.

Most of the plants are affected by diseases. To control the issue, it is necessary to know the status and condition of the plants/crops/trees instantly. Based on the plant's condition, precautionary steps can be taken to maintain high efficiency and protect the plants from viruses and diseases. With the ongoing revolution in the industry sector, digital twin technology plays a significant role in the 4th industrial revolution. Therefore, the proposed method involves using "Digital Twin" technology to monitor the conditions and status of the plants and crops. This can be achieved by creating a virtual representation of all the crops in the farmland [4].

In Digital Twin technologies, a simulation is created for the farmland and the crops to gather data such as water content in the soil, temperature, plant health, growth, and more. The proposed simulation includes adding external features like AI to continuously monitor the crops (24/7) and provide real-time updates on the status of the plants and crops. This method can also integrate the Internet of Things (IoT) to collect real-time information for testing and investigating various steps, as shown in Figure

11.2. Additionally, it is important to incorporate machine learning technology and factors like big data for the virtual models and simulations.

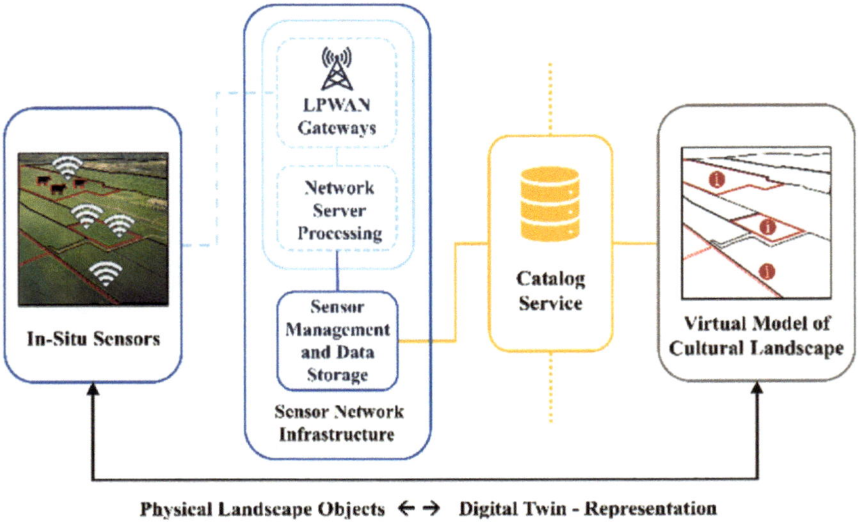

FIGURE 11.2 Landscape objects to digital representation.

For modeling and training, the proposed simulation needs some deep-learning concepts. All these technologies occupy more memory space. To overcome this specific disadvantage, it is planned to process the data and records into another form, i.e., "Blockchain." Blockchain is known as distributed ledger technology. It creates a database of digital assets that are chained together and stored for processing. It is decentralized, and real-time data can be accessed. Blockchain technology is encrypted and immutable. It is used in many applications like healthcare, gaming, and industries. In the proposed field, it is used for computing and storing transactions and records. Therefore, it improves the efficiency in the transmission of data at a higher speed.

11.2 LITERATURE REVIEW

In Ref. [5], the future challenges and trends in Blockchain-based digital twins have been discussed. Industries mostly prefer sensors to collect data. In regard to that, different types of sensors are used to collect different data from different areas so the data should be cross-validated to maintain high-quality data. It can be stored through blockchain to maintain data safely. Digital twin is used to identify faults and errors to give precautionary measures to recover the faults. It also describes the current research, development, and future challenges in the particular field.

In Ref. [6], they presented a review paper on digital twin applications in greenhouse horticulture. Many papers focused more on the cultivation process, energy management, and climate control. Specifically, unmanned horticulture is introduced using the digital twin in the proposed review paper. IoT along with the digital twin is used to make automation. The gained performance metrics using the predictive and prescriptive applications are used at their early stage itself. In Ref. [7], a review paper based on the espousal of digital twins in civil engineering was proposed. The drivers are identified for digital twin adoption in the civil engineering field and classified. The essential need for the digital twin in civil engineering business was emphasized. Therefore, it is aimed to execute an acceptance model which is a combination of mixed reality and Digital Twin MRDT.

In Ref. [8], they proposed a solution for farmers to solve environmental factors, planning, and task management using IoT. Agriculture faces great challenges in supplying substantial resources for the increasing number of residents around the world. In order to address the proposed issue, productivity needs to be increased. The proposed paper presented a classical procedure and strategic scheme for decision-making in intelligent farming. In Ref. [9], they presented a model to introduce digital twins in sustainable agriculture. Through IoT, productivity, self-protection, energy saving, and automation based on environmental changes are improved. The digital twin contributes by monitoring, planning, operating, and optimizing. In Ref. [10], they proposed a design based on IoT and Wireless Sensor Network (WSN) for monitoring the greenhouse.

In Ref. [11], a digital twin was developed for a vegetable crop. They proposed a multi-agent technology that predicts yield, vegetation quality, forecasts, plant growth, and timing for onset. The particular paper proposed a system based on ontologies and a multi-agent scheme to calculate the parameters. It mainly focused on the digital twin of a plant with agronomist work. In Ref. [12], the research established an idea of automation in a greenhouse environment using IoT. By using cameras, they collect data and monitor the water irrigation system and greenhouse environmental conditions.

In Ref. [13], they proposed a method of digital twin implementation for the cultivation of a field by sensing the changes and preserving water. A soil probe was developed for sensing the parameters and managing the irrigation process through the Internet of Things was established in the farm field. In the proposed paper, digital twin and blockchain technology were used to enhance the productivity, quality, and growth in agriculture based on analysis. Therefore, artificial intelligence, big data, IoT, and machine learning (processing the data) are extracted, and the analysis is carried out to produce better output in the farm field, which increases efficiency.

11.2.1 PROBLEM STATEMENT

Many countries' development depends on agriculture. Increasing world population and estimations show that in 2050 the population will be approximately 9 billion. Due to urbanization, people are moving to cities, specifically, 65% of people are residing in cities. It causes challenges in shelter and food production. Not only is the population increasing, but also the availability of areas is not. Due to this fact, agriculture needs to be improved in terms of production with fewer resources. The other main problem in agriculture is crops that are damaged due to diseases and insects. Even though insects help to maintain soil structure, fertility, and nutrients, pathogenic organisms like bacteria, fungi, viruses, and protozoa cause diseases in plants. This results in an average of 40% loss in agricultural production.

To overcome the problem, an idea of Digital twin technology is planned to monitor the farmland and crops. It enables the simulation models in the digital world to reflect in the physical world. Using this particular technology, the efficiency and growth of agriculture can be increased. In the proposed method, AI, IoT, and machine learning are integrated, requiring more memory space. To reduce the memory space, blockchain technology is used. Therefore, the data and records are encrypted.

11.3 PROPOSED METHODOLOGY

Digital twin connects the virtual and real world by gathering real-time data from IoT sensors and cameras. Through this process, the physical product can be represented in the digital world. This technology has a significant impact on development, production, and processes. In agricultural fields, IoT sensors such as optical sensors, electrochemical sensors, moisture sensors, location sensors, and temperature sensors are used to create a database. Additionally, the data may be stored either in a decentralized manner or centrally in the cloud. The collected data is processed using big data techniques. Subsequently, the data is tested and simulated in digital assets. Once the optimal solution is obtained from the simulation, it is applied to crops.

It helps in optimizing the performance of crops and plants. Digitalization of digital twins along with AI is more effective. Specifically, AI is employed to select the simulation options that have the highest chances so that AI reduces the amount of simulations to evaluate and also to store. Blockchain contains the records of the data from simulations, which are then applied to crops based on the analysis of data. Finally, the digital twin connects all the data collected in each technology together into a virtual model as shown in Figure 11.3.

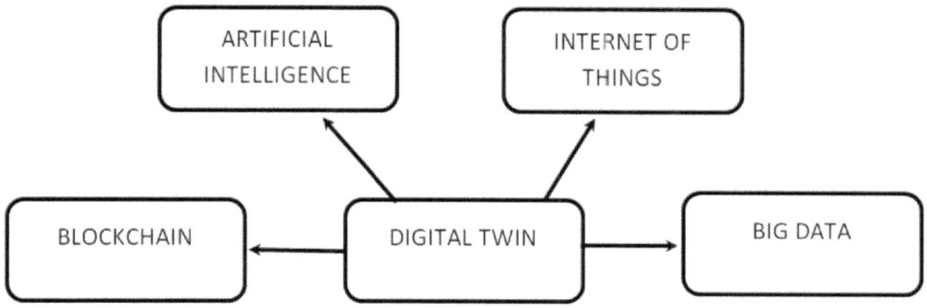

FIGURE 11.3 Connectivity of digital twin to the other technologies.

11.3.1 WORKING

In the proposed process, two places exist. One is the field area. In that, plants, crops, and some gadgets are available to obtain real-time data. The second part is the simulation center. In that, the current ongoing process can be viewed, along with previous actions and details of past data stored in a specific space. The future process of the crops can also be predicted. In simple terms, real-time input is collected from the field area, transmitted to the simulation center, processed, and stored as shown in Figure 11.4.

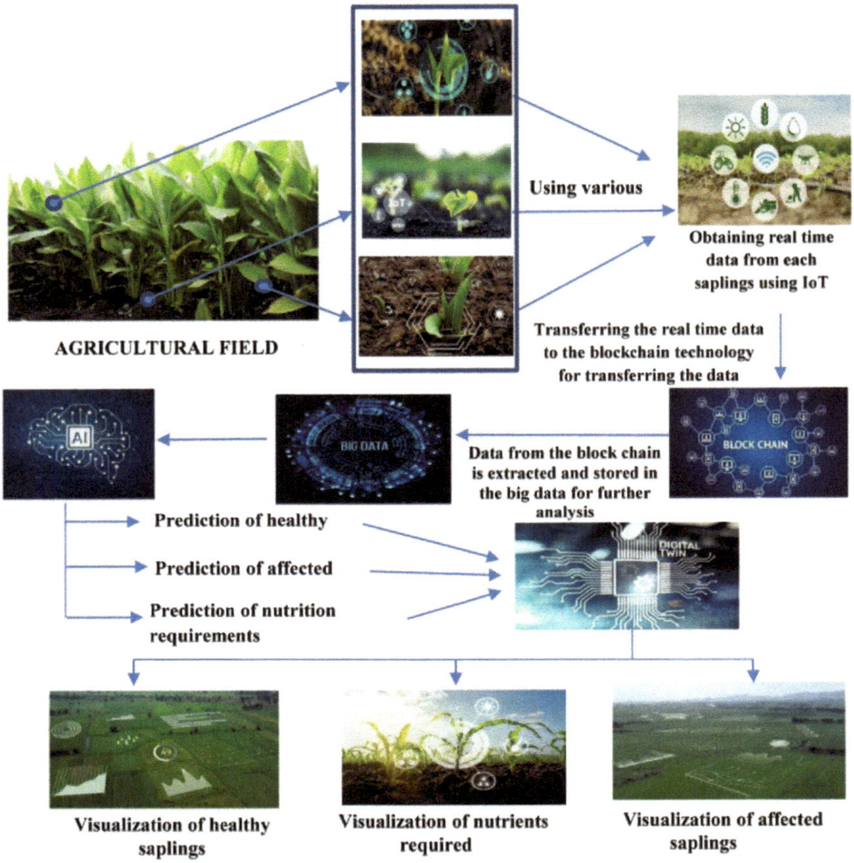

FIGURE 11.4 The step-by-step procedural flow of a process for making into a virtual model.

11.3.2 IoT

IoT is used to gather real-time data for creating a simulation [14–16]. In the proposed process, IoT stands for Internet of Things. It is a technology that uses sensors to convert real-time data (analog) into machine-level data (digital). The proposed technology mainly involves embedded system sensors, software, and other technologies (such as memory devices, displays, etc.). For the proposed simulation, five sensors and a camera for image processing are required: a sensor to detect soil moisture, a sensor to measure air humidity, a sensor to monitor soil temperature, a sensor to determine crop location and height, a sensor to measure groundwater level, and a camera for image processing. The specific image processing falls under deep learning:

1. **Moisture Sensor:** The proposed sensor is used to measure the humidity content of the soil. The default range is 0%RH–100%RH. It helps in reducing the water used for irrigation. If the moisture content in the soil is adequate, it stops the irrigation because, during that time, the crop doesn't need water.
2. **Temperature Sensor:** It is used to measure the temperature around the farmland. The default range is –40 to +80°C. It monitors the presence of heat in the soil. It helps to monitor whether the plant is in need of dry or wet conditions.
3. **Optical Sensor:** It is used to measure the nitrogen rate. It collects data on the condition of clay, organic matter, and humidity properties of the soil.
4. **Electrochemical Sensor:** These monitor the pH level and nutrient levels. Nutrients are essential for the plant to grow. By using the proposed sensor, nutrients can be provided based on the level.
5. **Location Sensor:** These are positioned at different places throughout the farm field. They are used to measure the variety, height, and distance of the crop plants. They also receive information from Global Positioning System (GPS) satellites.

As shown in Figure 11.5, using these sensors, it is decided to fetch data and carry out the process as per the data. Because of that, water can be saved. Using the wetness of soil, air, and groundwater levels, the water can be utilized as required so that it will not be wasted. In IoT, microcontrollers are the brain of the system [17]. So, in the proposed method, a microcontroller is used to collect the entire output of the sensor (raw data) and process the obtained data (convert into readable form) to convert it effectively by using the microcontroller's communication ports universal synchronous/asynchronous receiver/transmitter (USART), serial communication, parallel port communication, Inter-integrator circuits (I2C), etc. The data can be transferred to the blockchain module (which will be discussed later). Some more features such as controlling the water lines, controlling the animals' detection alarm, and so on can be added. Using the microcontroller's communication ports, the data from the simulation center to the microcontroller can easily be accessed. Also, using the particular communication, the water pipelines, and some more hardware devices are controlled as well. Each individual crop sapling data is sent to the blockchain.

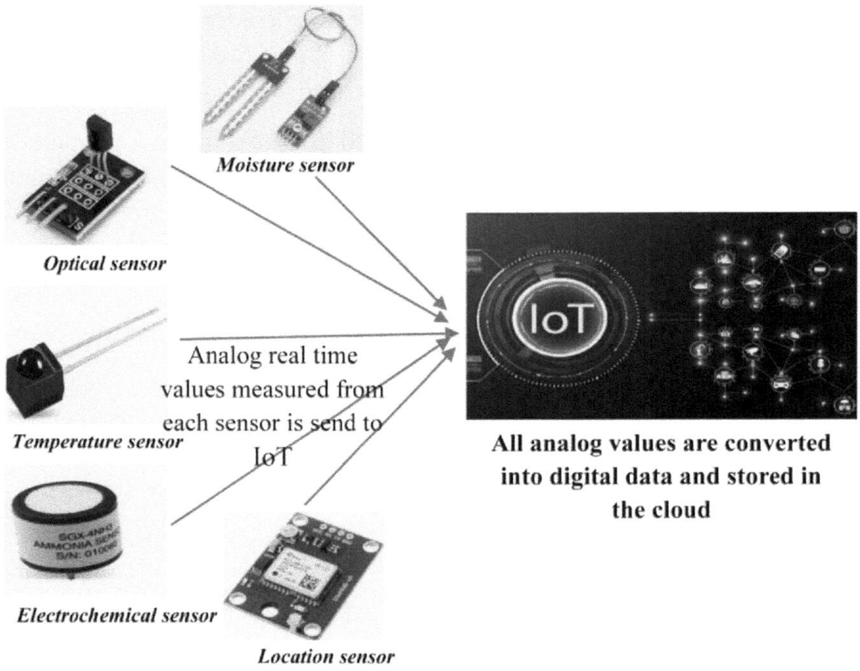

FIGURE 11.5 Measuring the soil parameters using various sensors with IoT.

11.4 BLOCKCHAIN

Blockchain is a new trend which is called distributed ledger technology [18]. It has a ledger for every input and communication so that the proposed blockchain is used. Actually, there are more field areas and because of that, all the data taken by the sensors is stored in clouds and transferred to the simulation center. Using the microcontroller itself, communication can be done to the simulation center, but it might be very slow, and the data may or may not be affected by the noise. Because of that, it may lead to the biggest error in the prediction. That is why Blockchain technology is proposed.

Using the particular method, communication can be done at high speed and the data in the blockchain is immutable. It is not able to update or insert any blocks into the chain, and it has high security, i.e., it is very safe to use for communication and storage of data. Moreover, it is generally used in big industries, banks, and mainly in the transaction of Bitcoins. As said before, the proposed module will get the information from the microcontroller (the processed raw data of the sensors) convert it into the respective format, and transmit the data to the simulation center.

In the simulation center, there is another blockchain module that will receive the information transmitted from the field area. It processes the data and sends it to the simulation. If any command is made by the AI, it is also transmitted from the blockchain module from the simulation center to the field area. That specific information is converted into data that will be given to the microcontroller as stated in the microcontroller block. It involves using particular technology for the proposed condition.

The first key point is "Distributed ledger technology," which means it has some ledger data containing information. This makes it very convenient for the customer using it. It also has smart contracts, connecting the blocks in the shortest way to access those blocks. It has an irreversible data structure, and each data block is connected to the one before and after it (like a doubly linked list). These are the main features of the proposed blockchain.

11.5 BIG DATA

Big data is one of the most famous ways to process raw data [19]. In the process, structured, semi-structured, and unstructured data are processed. Using this, the process can be sped up. This is only based on the user and customer convenience and comfort. It can be characterized by three 'V's (alphabetical letter, 22nd letter.) and they are as follows:

- The size of the data in every environment is very large.
- As big data is used to store huge data, a wide range and different types of data types are stored in the system.
- Data is stored in large amounts; then it looks into the velocity (speed) at which every data is being received and processed for storage.

In general, it is very useful for companies for marketing, advancement, insight into the customer, and so on. It is also used in the medical field for identifying diseases, using some signs (like the structure of a wound, color of the wound, etc.), finding the root of the diseases, and predicting risk factors. In the proposed work, there is a lot of data. Not all of it is from the same field area; data from different field areas and different sensors are available. It is important not only to write the data but also to read all the data in a faster manner, so it is better to opt for Big Data. In simple words, the sole purpose of the proposed technology is to store and access all the data in a faster and more comfortable manner.

11.6 AI

AI is used for decision-making. Actually, by using machine learning, the model can be trained. When compared to AI, the efficiency is low in machine learning. AI will think like humans, so using the technology, decisions can be made based on real conditions because AI is more efficient than humans [20]. That is the reason why it is planned to implement AI in the proposed work. In the current study, there are many simulation options available. Therefore, it is better to make decisions based on information to find the best way. Thus, the accuracy of the simulation can be precise as it mimics the real world. Making some changes can improve the efficiency of plants and crops. This cannot be achieved by human beings as it needs to be done quickly like machines, which is impossible. Programming is also challenging as predicting future conditions is a complex task.

There are nearly a trillion possible combinations available, which is why it is preferable to utilize AI to work with it. The main role of AI is to extract information from big data and process it [21]. It can identify efficient ways to maximize the quantity and quality of crops. Subsequently, it provides this information to the user (customer). By using real-time data and artificial intelligence, it is possible to determine the exact requirements for plants and crops. For instance, predicting the moisture content in the air and soil and sunlight intensity can help determine the water requirements.

If drip irrigation is used (it is one of the best irrigation types), it will transport the water to the roots of the plants so we can control the waste of water. Also, by using AI, it is easy to find out the required nutrients for the plants [22]. By doing so, the usage of fertilizer and some chemicals can be minimized. With this technology, fruits and vegetables can be protected from chemicals, and the quality of the produce can be maintained. The health of crops and nutrient content can be predicted. Unhealthy or disease-affected crops can be identified, and the level of disease spread as well as additional nutrient content required for crop growth improvement can be estimated. This data is then fed to the digital twin for visualization.

11.7 DIGITAL TWIN

A digital twin is a know-how system that brings up the imagination of real-time products into visualization with the support of factual information in the virtual environment [23]. It virtually shows the structure of an object with its interior parts. It helps the developers to see their product virtually and make changes or get alerted before they create it into a model. The handling of the components and its simulation process is made easier by digital twins. So, as in agriculture, the agricultural fields are being shown virtually by digital twin using the factual information extracted from the previous process.

The information received from the AI prediction is made into 3D visualization. The agricultural field is made into a virtual model and in that, the affected crops are being located by the location data obtained from the sensor. On looking into the particular location, the 3D visualization of the crop is made visible [24]. On the right side of the virtual model, the predicted data is shown for reference (disease name, disease spread percentage, nutrient content, required maintenance, etc.). On the other hand, the estimation of healthy species and their levels are displayed.

As shown in Figures 11.6 and 11.7, by using digital twin technology, field management is made easier, allowing farmers to identify diseased crops accurately and at an earlier stage. This helps satisfy the further requirements for the crops to recover from diseases. The virtual model of each crop can be viewed with options for rotation and zooming, making analysis more flexible and faster. As a result, the crop yield percentage increases, and due to effective management, the nutrient content remains rich.

FIGURE 11.6 A virtual model of a healthy crop.

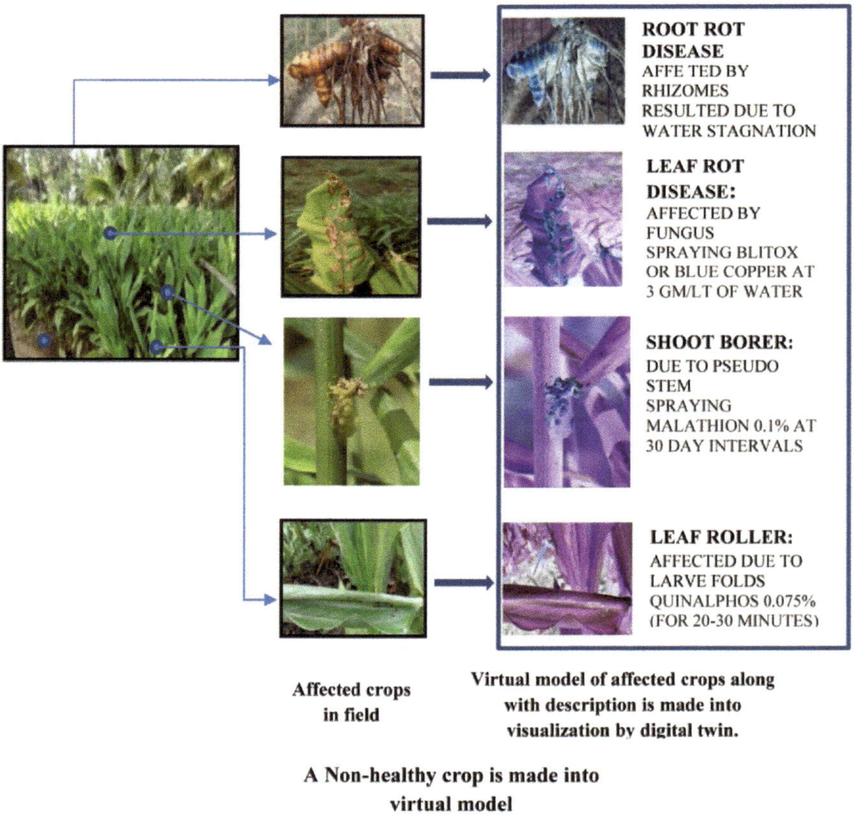

FIGURE 11.7 A virtual model of non-healthy crops.

11.8 CONCLUSION

Using the proposed technology, all the resources needed for the crops can be reduced. Moreover, it reduces waste by increasing the efficiency of the crops' productivity, reduces the infected crops due to diseases or viruses, reduces investments in the field, predicts future progress so that it can be planned accordingly, reduces the manual work in the field, increases the quality of the crops, fruits, and vegetables, and also maintains the quality of the soil. These are the advantages of the proposed method. The only problem here is data or information maintenance and security. By using big data, the issue can be nullified. It is also better to use the cloud to store all the data because it is closer to the 6th generation (6G). Due to that, the cloud has faster accessibility. The data and records will be safely stored in Blockchain.

KEYWORDS

- **agriculture field**
- **blockchain technology**
- **digital twin**
- **internet of things**
- **non-healthy crops**
- **sensors**
- **small-scale productions**

REFERENCES

1. Chaux, J. D., Sanchez-Londono, D., & Barbieri, G. (2021). A digital twin architecture to optimize productivity within controlled environment agriculture. *Applied Sciences, 11*(19), 8875.
2. Nasirahmadi, A., & Hensel, O. (2022). Toward the next generation of digitalization in agriculture based on the digital twin paradigm. *Sensors, 22*(2), 498.
3. Maheswaran, S., Sathesh, S., Rithika, P., Shafiq, I. M., & Nandita, S. (2022). Detection and Classification of Paddy Leaf Diseases Using Deep Learning (CNN).

In *Detection and Classification of Paddy Leaf Diseases Using Deep Learning (CNN)* (pp. 60–74). Springer.
4. Pylianidis, C., Osinga, S., & Athanasiadis, I. N. (2021). Introducing digital twins to agriculture. *Computers and Electronics in Agriculture, 184*, 105942.
5. Suhail, S., Hussain, R., Jurdak, R., & Oracevic, A., & Salah, K. (2022). Blockchain-based digital twins: Research trends, issues, and future challenges. *ACM Computing Surveys (CSUR), 54*, 1–34.
6. Ariesen-Verschuur, N., Verdouw, C., & Tekinerdogan, B. (2022). Digital Twins in greenhouse horticulture: A review. *Computers and Electronics in Agriculture, 199*, 107183.
7. Sepasgozar, S. M., Ghobadi, M., Shirowzhan, S., Edwards, D. J., & Delzendeh, E. (2021). Metrics development and modeling the mixed reality and digital twin adoption in the context of Industry 4.0. *Engineering, Construction, and Architectural Management, 28*(5), 1355–1376.
8. Suakanto, S., Engel, V. J., Hutagalung, M., & Angela, D. (2016). Sensor networks data acquisition and task management for decision support of smart farming. In *Sensor Networks Data Acquisition and Task Management for Decision Support of Smart Farming* (pp. 1–5). IEEE.
9. Monteiro, J., Barata, J., Veloso, M., Veloso, L., & Nunes, J. (2018). Towards sustainable digital twins for vertical farming. In *Towards Sustainable Digital Twins for Vertical Farming* (pp. 234–239). IEEE.
10. Touhami, A., Benahmed, K., & Bounaama, F. (2020). Monitoring of Greenhouse Based on the Internet of Things and Wireless Sensor Network. In *Monitoring of Greenhouse Based on Internet of Things and Wireless Sensor Network* (pp. 281–289). Springer.
11. Skobelev, P., Laryukhin, V., Simonova, E., Goryanin, O., & Yalovenko, V. (2020). Multi-agent approach for developing a digital twin of wheat. In *Multi-Agent Approach for Developing a Digital Twin of Wheat* (pp. 268–273). IEEE.
12. Khan, F. A., Ibrahim, A. A., & Zeki, A. M. (2020). Environmental monitoring and disease detection of plants in a smart greenhouse using the internet of things. *Journal of Physics Communications, 4*(5), 055008.
13. Alves, R. G., Souza, G., Maia, R. F., Tran, A. L. H., & Kamienski, C. (2019). A digital twin for smart farming. In *A Digital Twin for Smart Farming* (pp. 1–4). IEEE.
14. Muneeswari, G., Sajithra, S., & Hegde, R. (2023). Self-diagnosis platform via IoT-based privacy preserving medical data. *Measurement: Sensors, 25*, 1–10.
15. Dhiviya, S., Malathy, S., & Kumar, D. R. (2018). Internet of Things (IoT) Elements, Trends and Applications. *Journal of Computational and Theoretical Nanoscience, 15*, 1639–1643.
16. Vazhuthi, P., Aruchamy, P., & Manikandan, S. P. (2023). A hybrid ANFIS reptile optimization algorithm for energy-efficient inter-cluster routing in Internet of Things-enabled Wireless Sensor Networks. *Peer-to-Peer Networking and Applications, 16*, 1049–1068.
17. Shanmugam, M., & Ramasamy, A. (2014). Sensor-based turmeric finger growth characteristics monitoring using embedded systems under the soil. *International Journal of Distributed Sensor Networks, 10*(6), 476176.

18. Wang, K., Xie, W., Wang, B., Pei, J., & Wu, W. (2020). Simulation-based digital twin development for blockchain-enabled end-to-end industrial hemp supply chain risk management. In *Simulation-Based Digital Twin Development for Blockchain-Enabled End-To-End Industrial Hemp Supply Chain Risk Management* (pp. 3200–3211). IEEE.
19. Pushpalatha, N., Sudha, M., & Sivaranjani, S. (2023). Enterprise Data, Data Fabric Architectures: Web-Driven Applications (pp. 95–116).
20. Maheswaran, S., Indhumathi, N., Dhanalakshmi, S., Nandita, S., & Mohammed Shafiq, I. (2022). Identification and Classification of Groundnut Leaf Disease Using Convolutional Neural Network. In *Identification and Classification of Groundnut Leaf Disease Using Convolutional Neural Network* (pp. 251–270). Springer.
21. Kavitha, M., & Roobini, S. (2023). Systematic View and Impact of Artificial Intelligence in Smart Healthcare Systems, Principles, Challenges and Applications. In *Machine Learning and Artificial Intelligence in Healthcare Systems* (pp. 25–56).
22. Maheswaran, S., Sathesh, S., Priyadharshini, P., & Vivek, B. (2017). Identification of artificially ripened fruits using smartphones. In *Identification of artificially ripened fruits using smartphones* (pp. 1–6). IEEE.
23. Purcell, W., & Neubauer, T. (2022). Digital twins in agriculture: A state-of-the-art review. *Smart Agricultural Technology,* 100094.
24. Angin, P., Anisi, M. H., Göksel, F., Gürsoy, C., & Büyükgülcü, A. (2020). AgriLoRa: A digital twin framework for smart agriculture. *Journal of Wireless Mobile Networks, Ubiquitous Computing, and Dependable Applications, 11*(4), 77–96.

Index